Letts study aids

Objective Questions Chemistry

for O Level, CSE and Scottish O Grade

G. R. McDuell BSc (Hons)

Deputy Headmaster, Berry Hill High School, Stoke-on-Trent

Charles Letts & Co Ltd
London, Edinburgh & New York

Preface

This book has been written as a companion to the *Revise Chemistry* O-level and CSE guide. Together, they provide you with a complete preparation for your Chemistry examination. We have received many letters of appreciation from students who have found the style and content of *Revise Chemistry* extremely helpful in preparation for examinations. Also, teachers have welcomed it as a most useful revision aid.

It would be nice to think that you could prepare for your examination with crosswords and puzzles but it is well-understood that the best method of preparing for an examination involves the study of suitable examination-type questions. It is impossible to attempt too many questions and this book is designed to provide a large number of suitable questions to assist with your examination preparation.

This book contains over seven hundred fixed response (or multiple choice) questions of all types. Candidates often find these questions difficult because they highlight common problems and difficulties. The questions in this book are especially written and tested to provide you with a wide range of typical O-level and CSE questions. There are sufficient fixed response questions in this book to set examination papers for the next twenty years!

In addition, there are a number of longer structured questions. A structured question consists of a number of questions about a common situation or set of facts. The structured questions emphasize the common errors that are made and how they can be avoided. Over the past ten years the nature of structured questions has changed. There is evidence now that they provide more information than they previously did and require more skills of comprehension for their solution. The questions in this book reflect these changes and a number of comprehension-type questions are also included.

I would like to thank Dr Krysia Ahmad for her assistance in the preparation of this book and the staff of Charles Letts Books Ltd for the efficient transformation of a manuscript into a book. Again, I must thank my wife, Judy, and sons Robin and Timothy for their help during the writing of this book.

Bob McDuell 1983

Guide to using this book

When using this book you should isolate the units that are required for your particular syllabus. No syllabus requires you to study all of this book. Remember, in *Revise Chemistry* pages 2 and 3, there is an extremely useful table listing the syllabus requirements for every examination syllabus. In this book you will find a table on page 41 listing the types of question set on each examination syllabus. Having isolated the important units, you will find that questions in this book are related to the units in *Revise Chemistry*. If you are required to study unit 33 in *Revise Chemistry,* for example, you will find that this book contains 34 fixed response questions and one structured question on this unit.

Before attempting the questions on a particular unit you should revise the topic thoroughly by reading the appropriate unit in *Revise Chemistry* and attempting the self-test or using your school or college notes. Then work through the questions. It is important that you do not take too long over fixed response questions. Allow yourself a maximum of one minute per fixed response item. Having worked through the questions check your answers against the correct answers on pages 69–76. Return immediately to the corresponding unit in *Revise Chemistry* or your notes to find out why you have made any mistakes.

First published 1983
by Charles Letts & Co Ltd
Diary House, Borough Road, London SE1 1DW

Reprinted 1983

Editor: Krysia Ahmad
Design: Ben Sands, Anne Davison
Illustrations: Rod Fraser

ISBN 0 85097 585 9

Printed in Great Britain by Charles Letts (Scotland) Ltd

Contents

Table of analysis of question types of different boards

Board		Fixed Response Simple	Classification	Multiple completion	Assertion–reason	Free Response Short answer	Structured	Essay or Extended answer
SEB		●	●			●		●
AEB	1	●	●	●		●		●
	2					●	●	●
Cambridge		●	●	●				●
JMB		●	●				●	●
London		●	●	●	●		●	●
London Nuffield`		●	●	●	●		●	●
NIEC		●	●		●		●	●
Oxford	5855					●	●	●
	5856					●	●	●
O & C		●	●	●			●	●
SUJB						●	●	●
WJEC						●	●	●
ALSEB		●				●	●	
EAEB	**Joint**					●	●	
	Nuffield	●	●				●	
EMREB	1					●	●	
	2	●	●	●		●	●	
LREB	A	●				●	●	●
	B	●	●			●	●	
	C	●				●	●	●
	D	●	●			●	●	
NREB						●	●	
NWREB		●				●	●	●
SREB	A	●	●				●	
	B	●	●				●	
SEREB	Traditional	●	●	●			●	
	Nuffield	●	●	●			●	
SWEB		●				●	●	
WJEC (CSE)						●	●	●
WMEB	A	●	●			●	●	
	B	●	●			●	●	
YHREB (former WY & LREB syllabus)						●	●	●
YHREB (former LREB syllabus)						●	●	
JMB/WMEB 16+		●	●	●	●	●	●	

Analysis of question types used by different examination boards

The table on page iv outlines the different types of questions used by the various examining boards.

Fixed response questions

Fixed response questions are sometimes called multiple choice or objective questions. These questions can be divided into various types.

1 Simple multiple choice

In this type of item an incomplete statement or question (called the stem) is given together with four or five possible answers (or responses) and the student has to choose the only answer that correctly fits the stem. The wrong answers are called distractors.

e.g. What colour is hydrated copper(II) sulphate?

A blue B green C white D yellow E black

A is the correct answer (or key). B, C, D and E are distractors. This question is entirely recall of information that the candidate should have learnt. If she confuses hydrated copper(II) sulphate with anhydrous copper(II) sulphate, she would answer C.

2 Classification

This type of question is used when a series of similar simple multiple questions are set using identical responses.

e.g. For each of the questions **1–4** choose the one process labelled A, B, C, D or E with which it is chiefly associated.

A cracking B polymerization C oxidation D hydrogenation E neutralization

1 Decane vapour is changed to ethene when passed over heated china.
2 Margarine is produced by passing hydrogen through a heated oil in the presence of a catalyst.
3 Sodium carbonate is added to ethanoic acid (acetic acid) until no further carbon dioxide is evolved.
4 Ethene is completely burnt in excess oxygen.

The correct responses to these questions are:

1 A. **2** D. **3** E. **4** C.

In this type of question each response may be used once, more than once or not at all.

3 Multiple completion questions

Basically these questions use a stem together with either three or four responses. One or more of these responses is (are) correct. Different examination boards use different combinations of responses. For the purpose of this book they are subdivided into type 1 and type 2. Type 1 is probably typical of a CSE paper and Type 2 is typical of an O-level paper.

Type 1

In each question one or more of the given answers is (are) correct. Decide which one or ones are correct, then write your answer using the following code.

A If (*i*), (*ii*) and (*iii*) are correct.
B If (*i*) and (*ii*) are correct.
C If (*ii*) and (*iii*) are correct.
D If (*i*) only is correct.
E If (*iii*) only is correct.

Type 2

In each question one or more of these responses is (are) correct.
Choose

A If only (*i*), (*ii*) and (*iii*) are correct.
B If only (*i*) and (*iii*) are correct.
C If only (*ii*) and (*iv*) are correct.
D If only (*iv*) is correct.
E If some other response, or combination of responses is (are) correct.

These can be summarized as follows:

A	B	C	D	E
(*i*), (*ii*), (*iii*) only	(*i*), (*iii*) only	(*ii*), (*iv*) only	(*iv*) only	Some other response or combination of responses.

4 Assertion-reason

Each of the questions consists of a statement in the left-hand column followed by a second statement in the right-hand column.

Decide whether the first statement is true or false.
Decide whether the second statement is true or false.

Then choose:

A If both statements are true and the second statement is a correct explanation of the first statement.

B If both statements are true but the second statement is **not** a correct explanation of the first statement.

C If the first statement is true, but the second statement is false.

D If the first statement is false, but the second statement is true.

E If both statements are false.

These can be summarized as follows:

	First statement	*Second statement*	
A	True	True	Second statement is a correct explanation of the first.
B	True	True	Second statement is **not** a correct explanation of the first.
C	True	False	
D	False	True	
E	False	False	

1 Elements, mixtures and compounds

Simple multiple choice

1 Which one of the following statements is the **best** description of an element?
An element is a substance which

A can be a metal or a non metal.
B is not a compound.
C cannot be split into simpler substances.
D is composed of atoms.

2 Which one of the substances in bold type is an element?

A a **green powder** which turns black on heating and evolves a colourless gas.
B a **colourless liquid** which boils at 58°C.
C a **colourless liquid** which is split up by electricity forming two colourless gases.
D a **dark-grey powder** which burns in oxygen to form a single colourless gas.

3 Five samples of the element sulphur were burnt in oxygen and the sulphur dioxide produced was weighed. Which one of the samples of sulphur was impure?

	Mass of sulphur	*Mass of sulphur dioxide*
A	0.50 g	1.00 g
B	0.64 g	1.28 g
C	0.75 g	1.32 g
D	0.80 g	1.60 g
E	0.85 g	1.70 g

4 Which one of the following lists consists of three mixtures?

A air, crude oil, salt solution
B water, carbon dioxide, copper
C soil, salt, liquid air
D hydrogen, steel, glass

5 Which one of the following statements is the **best** description of a chemical compound? A chemical compound is a substance which

A contains only one type of atom.
B is a mixture of two or more elements.
C is the product of a chemical reaction.
D contains two or more elements that are chemically combined.
E contains fixed quantities of two or more elements.

6 When two elements X and Y combine to form a compound XY, the properties of XY are

A similar to the properties of X.
B similar to the properties of Y.
C a mixture of the properties of X and Y.
D completely different from the properties of either X or Y.

7 Which one of the following is a single pure compound?

A air B steam C brass D sea water

8 Which one of the following compounds is composed of only two elements?

A sodium hydroxide
B sodium nitrate
C sodium carbonate
D sodium sulphate
E sodium hydride

9 Which one of the following compounds contains more than two elements?

A zinc sulphate
B hydrogen chloride
C sulphur trioxide
D aluminium sulphide
E calcium oxide

10 When electricity is passed through water, hydrogen and oxygen are collected in the ratio of 2 : 1 by volume. This shows that water

A is a mixture of hydrogen and oxygen.
B contains dissolved hydrogen and oxygen.
C is a chemical element.
D is a chemical compound.

Classification

Questions 11–15
Substances may be classified into groups as follows:

A single element.
B mixture of elements.
C single compound.
D mixture of compounds.
E mixture of elements and compounds.

For each of the following substances, select the group A–E to which they belong.
11 air **12** solder **13** pure water **14** milk **15** tin

Multiple completion

Type 1 (See instructions on page 1.)

16 The properties of a pure compound are the same
(*i*) as the properties of its constituent elements.
(*ii*) as the properties of other pure compounds containing the same elements.
(*iii*) whichever method is used for the preparation of the compound.

17 Which of the following equations represent examples of synthesis?
(*i*) $N_2(g) + 3H_2(g) \rightleftharpoons 2NH_3(g)$
(*ii*) $2Na(s) + Cl_2(g) \rightarrow 2NaCl(s)$
(*iii*) $CuO(s) + H_2(g) \rightarrow Cu(s) + H_2O(l)$

18 A mixture of iron and sulphur can be distinguished from the compound iron(II) sulphide by using
(*i*) a magnet.
(*ii*) dilute hydrochloric acid.
(*iii*) heating.

Type 2 (See instructions on page 1.)

19 Which of the following compounds contain the elements sodium, sulphur and oxygen?
(*i*) sodium sulphite
(*ii*) sodium thiosulphate
(*iii*) sodium sulphate
(*iv*) sodium sulphide

Assertion-reason

(See instructions on pages 1–2.)

20 Calcium, carbon and oxygen are all elements.	Calcium carbonate is a compound composed of calcium, carbon and oxygen.
21 Elements can be split up by electricity.	Atoms can be split up into smaller particles.
22 Hydrated cobalt(II) chloride contains water.	Water is a compound of hydrogen and oxygen.

2 Separation techniques in chemistry

Simple multiple choice

1 In **any** process used to separate one substance from another, use is always made of
A a chemical reaction.
B a property which differs for the two substances.
C differences in melting and boiling points.
D a liquid which dissolves only one of the substances.

2 The best definition of a solute is
A a substance which is to be dissolved in a liquid.
B the liquid in which a substance is dissolved.
C the solid formed when two solutions are mixed.
D the solid obtained when a gas changes to a solid.

3 Chromatography is a process by which
A a metal is plated with chromium.
B coloured dyes are removed by bleaching.
C chemically similar substances in solution are separated.
D dyes are extracted from a plant using a solvent.

4 Oxygen is manufactured from liquid air by fractional distillation. This process depends upon the fact that
A oxygen is more reactive than nitrogen.
B nitrogen is cooled more quickly than oxygen.
C liquid oxygen is denser than liquid nitrogen.
D liquid nitrogen has a lower boiling point than liquid oxygen.

5 It is suspected that a company is using a dangerous yellow dye to colour an orange squash. Which one of the following methods is used to find out if orange squash contains this particular dye?
A centrifugation
B chromatography
C distillation
D filtration
E evaporation

Questions 6–9 refer to Fig. 2.1.

6 The apparatus in Fig. 2.1 can be used to extract propanone (boiling point 56°C) from a solution of plant dyes dissolved in propanone. The process taking place is best described as
A evaporation.
B distillation.
C chromatography.
D fractional distillation.
E condensation.

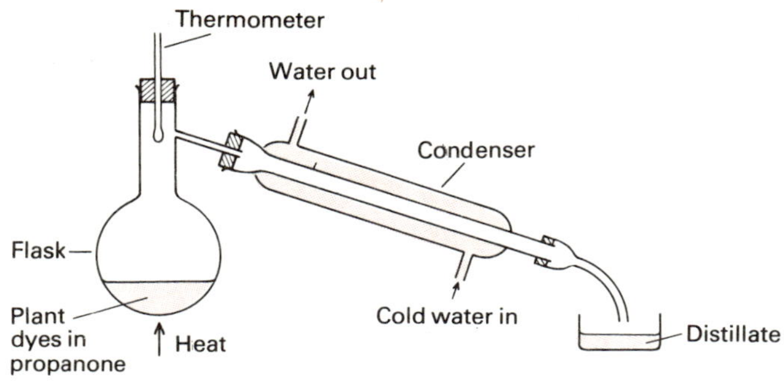

Fig. 2.1

7 The maximum temperature recorded on the thermometer during the process is approximately
A 20°C. B 55°C. C 85°C. D 100°C.

8 The temperature of the distillate collected is approximately
A 20°C. B 55°C. C 85°C. D 100°C.

9 During the process the plant dyes
A react to form colourless products.
B remain in the flask.
C are removed by the cooling water.
D are collected in the distillate.

Questions 10 and 11 refer to Fig. 2.2.

10 Figure 2.2 shows a method which can be used to purify some impure solids.

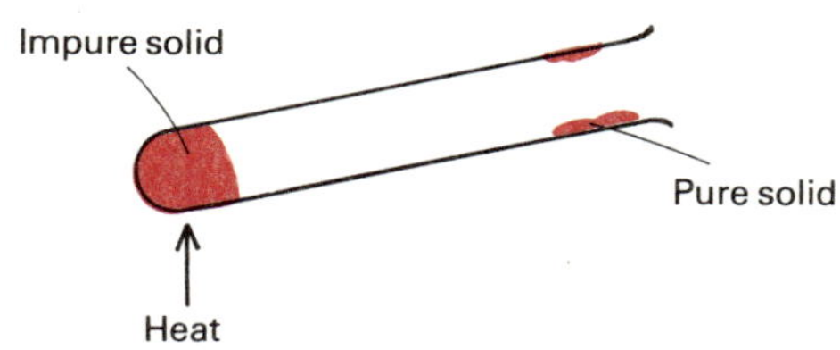

Fig. 2.2

For a substance to be suitable for purification by this method, it must
A be soluble in water.
B have a low relative molecular mass.
C have a high boiling point.
D react with air.
E be easily vaporized.

11 Which one of the four substances A–D in the table below could be separated using the apparatus in Fig. 2.2?

	Appearance	*Action of heat*
A	white solid	melts to form a colourless liquid
B	blue solid	turns white and evolves steam
C	white solid	turns yellow
D	white solid	turns to colourless gas. No residue

12 A tap funnel is most suitable for separating
A two powdered solids.
B an undissolved solid and a solution.
C two immiscible liquids.
D two miscible liquids.
E a solute from a solvent solution.

Classification

Questions 13–18

Here is a list of methods for purifying substances:

A chromatography B distillation C evaporation D filtration

Which method is most suitable for each of the following?

13 Removing small pieces of glass from a solution.
14 Obtaining a sample of salt from a salt solution.
15 Producing a nearly pure sample of ethanol from a sample of beer.
16 Separating the mixture of amino acids produced by hydrolysis of a protein.
17 Producing water for domestic use from a reservoir.
18 Producing a sample of pure water from sea water.

(You may use any answer once, more than once or not at all.)

Multiple completion

Type 1 (See instructions on page 1.)

19 For a pure substance
(*i*) the melting point is over a range of temperature.
(*ii*) the boiling point is increased by adding salt.
(*iii*) the boiling point is reduced when atmospheric pressure is lowered.

20 Pure sulphur can be obtained from a mixture of powdered iron and sulphur by
(*i*) removing the sulphur by dissolving in xylene.
(*ii*) using a magnet to remove the iron.
(*iii*) removing the iron by reaction with hydrochloric acid.

Type 2 (See instructions on page 1.)

21 A solution of substance A was boiled with dilute hydrochloric acid. A chromatography experiment was carried out using a drop of solution A before boiling and also a drop of the solution which remained after boiling. Solutions of three possible reaction products X, Y and Z were also used for comparison. The result is shown in Fig. 2.3.

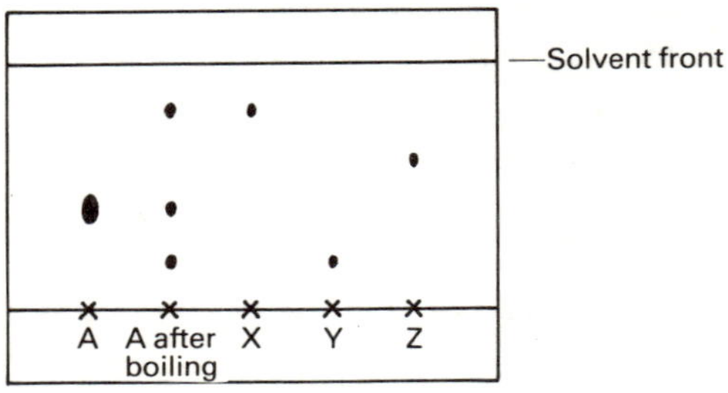

Fig. 2.3

Which of the following conclusions is (are) correct?
(*i*) Some of the substance A remained unreacted after boiling.
(*ii*) X is a product of the reaction of A with hydrochloric acid.
(*iii*) Y is a product of the reaction of A with hydrochloric acid.
(*iv*) Z is a product of the reaction of A with hydrochloric acid.

Assertion-reason

(See instructions on pages 1–2.)

	Assertion	Reason
22	Ethanol and water can be separated using a tap funnel.	Ethanol and water are immiscible.
23	Water can be used in the purification of rock salt.	Water is a good solvent for salt but not for the impurities in rock salt.
24	Liquid air can be separated into its component gases by fractional distillation.	Liquid air has the same composition as gaseous air.

Structured questions

25 An experiment was carried out with the dye extracted from elm tree leaves using propanone.

The leaves were put into a mortar with a little sand and some propanone. The mixture was then crushed with a pestle. The mixture was then added to a boiling tube and a further 10 cm^3 of propanone were added. The boiling tube and contents were heated to about 50°C.

(*a*) Why is sand added to the mortar?
(*b*) Propanone is very flammable. How could the boiling tube and its contents be heated safely?
(*c*) As the boiling tube and contents were heated, propanone evaporated away. Using the apparatus in the list below, plus any corks, rubber tubing and a source of heat, draw a diagram of the apparatus that could be used to heat the solution without propanone being lost by evaporation.
flask   condenser   flask with sidearm   beaker
(*d*) The remains of the leaves and the sand were then removed from the solution. What process can be used to bring about this separation?
(*e*) A drop of the dye solution from the leaves was placed in the centre of a piece of filter paper and allowed to dry. Then drops of propanone were added to the centre of the filter paper. The spot was enlarged. The green dye separated into two rings – a green ring due to chlorophyll and a yellow ring due to xanthrophyll.
(*i*) Draw a labelled diagram of the filter paper at the end of the experiment. (Chlorophyll does not travel across the filter paper as readily as xanthrophyll.)
(*ii*) How could minute samples of pure chlorophyll and pure xanthrophyll be obtained from this filter paper.

(*f*) A similar experiment was carried out using green leaves collected in Spring and some leaves which had fallen from the tree in Autumn. The chromatography was carried out with strips of filter paper and the results are shown in Fig. 2.4.

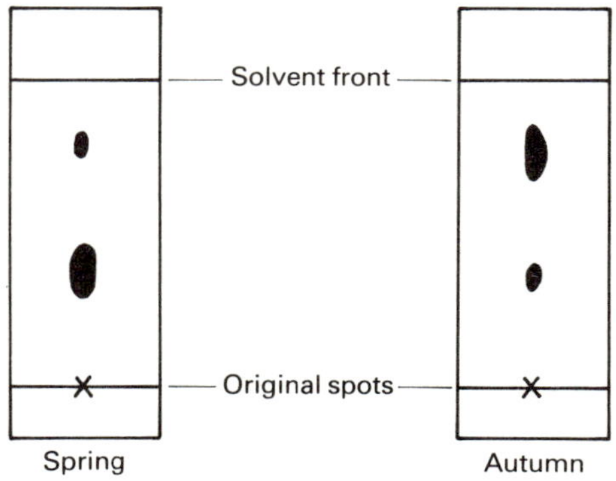

Fig. 2.4

What can be concluded about the dyes present in the leaves in Spring and Autumn?

(*g*) At the end of the experiment the teacher wanted to recover the propanone from the remaining dye solution. Using the apparatus listed in (*c*), draw a diagram of the apparatus set up to recover the propanone.

3 Symbols, formulae and equations

Simple multiple choice

1 $2H_2$ represents
A two hydrogen molecules.
B two hydrogen atoms.
C two hydrogen ions.
D four free hydrogen atoms.

2 In which one of the following lists are all the formulae correct?
A H_2SO_4, NH_4, $Cu(NO_3)_2$
B HNO_3, CH_4, Al_2O_3
C $MgSO_4$, HCl_2, KNO_3
D $NaCO_3$, H_2O, $Ca(OH)_2$
E $CuCl_2$, CO_3, MgO

3 A metal M forms ions with 1+, 2+ and 3+ charges. Which one of the following gives the correct formulae of the three possible nitrates of M?
A MNO_3, $M(NO_3)_2$, $M_2(NO_3)_3$
B M_2NO_3, MNO_3, $M(NO_3)_3$
C MNO_3, $M(NO_3)_2$, $M(NO_3)_3$
D M_2NO_3, $M_2(NO_3)_2$, $M_2(NO_3)_3$
E MNO_3, $M_2(NO_3)_2$, $M(NO_3)_3$

4 The bromide ion is Br^- and the formulae of calcium bromide and calcium phosphate are $CaBr_2$ and $Ca_3(PO_4)_2$ respectively. What is the charge on the phosphate ion?
A +3 B +2 C −1 D −2 E −3

5 The formula of a compound of the metal M with chlorine is MCl. What is the formula of the hydroxide of M?
A MOH B $M(OH)_2$ C $M(OH)_3$ D $M(OH)_4$ E M_2OH

6 Which one of the following equations correctly represents the burning of magnesium in oxygen?
A $Mg(s) + O_2(g) \rightarrow MgO_2(s)$
B $Mg(s) + 2O(g) \rightarrow MgO_2(s)$
C $2Mg(s) + O_2(g) \rightarrow 2MgO(s)$
D $4Mg(s) + O_2(g) \rightarrow 2Mg_2O(s)$
E $Mg(s) + O(g) \rightarrow MgO(s)$

7 Calcium dicarbide reacts with water to form ethyne gas (acetylene) and calcium hydroxide solid. The correct equation for this reaction is
A $CaC_2(s) + H_2O(l) \rightarrow CaO(s) + C_2H_2(g)$.
B $CaC_2(s) + 2H_2O(l) \rightarrow Ca(OH)_2(s) + C_2H_2(g)$.
C $CaC_2(s) + H_2O(l) \rightarrow CaOH(s) + C_2H_2(g)$.
D $CaC_2(s) + 2H_2O(l) \rightarrow Ca(OH)_2(aq) + 2CH(g)$.

8 Which one of the following equations is **not** correctly balanced?
A $Mg(s) + 2H^+(aq) \rightarrow Mg^{2+}(aq) + H_2(g)$
B $Ca(s) + 2H_2O(l) \rightarrow Ca(OH)_2(aq) + H_2(g)$
C $Fe^{2+}(aq) + Ag^+(aq) \rightarrow Fe^{3+}(aq) + Ag(s)$
D $Fe^{2+}(aq) + Cl_2(g) \rightarrow Fe^{3+}(aq) + 2Cl^-(aq)$

9 Which one of the following is the correct ionic equation for the reaction between zinc and dilute hydrochloric acid?

A $Zn(s) + 2HCl(aq) \rightarrow ZnCl_2(aq) + H_2(g)$
B $Zn(s) + 2HCl(aq) \rightarrow Zn^{2+}(aq) + 2Cl^-(aq) + H_2(g)$
C $Zn(s) + H^+(aq) \rightarrow Zn^+(aq) + \frac{1}{2}H_2(g)$
D $Zn(s) + 2H^+(aq) \rightarrow Zn^{2+}(aq) + 2H(g)$
E $Zn(s) + 2H^+(aq) \rightarrow Zn^{2+}(aq) + H_2(g)$

10 The equation for the reaction between copper and dilute nitric acid is
$x Cu + y HNO_3 \rightarrow 3Cu(NO_3)_2 + 4H_2O + 2NO$
Which one of the following gives values of x and y which will result in a balanced equation?

A $x=1, y=4$
B $x=2, y=3$
C $x=3, y=8$
D $x=4, y=7$
E $x=2, y=5$

11 The simplest ionic equation for the neutralization of hydrochloric acid with sodium hydroxide is

A $Na^+(aq) + Cl^-(aq) \rightarrow NaCl(aq)$.
B $NaOH(aq) + HCl(aq) \rightarrow NaCl(aq) + H_2O(l)$.
C $H_2(g) + \frac{1}{2}O_2(g) \rightarrow H_2O(l)$.
D $H^+(aq) + OH^-(aq) \rightarrow H_2O(l)$.

Classification

Questions 12–17

A a solution only
B a solution and a gas
C a precipitate and a solution
D a solid only
E a gas and a solid

In each of the questions **11–17,** select from the list above the best description of the products obtained.

12 $Zn(s) + CuSO_4(aq) \rightarrow Cu(s) + ZnSO_4(aq)$
13 $CaCO_3(s) \rightarrow CaO(s) + CO_2(g)$
14 $NaOH(aq) + HNO_3(aq) \rightarrow NaNO_3(aq) + H_2O(l)$
15 $Fe(s) + S(s) \rightarrow FeS(s)$
16 $CaCO_3(s) + 2HCl(aq) \rightarrow CaCl_2(aq) + H_2O(l) + CO_2(g)$
17 $ZnSO_4(aq) + Na_2CO_3(aq) \rightarrow ZnCO_3(s) + Na_2SO_4(aq)$

Multiple completion

Type 1 (See instructions on page 1.)

18 The table below summarizes some information about zinc carbonate and zinc oxide.

	Colour when hot	*Colour when cold*
Zinc carbonate	—	White
Zinc oxide	Yellow	White

The equation for the action of heat on zinc carbonate is
$ZnCO_3(s) \rightarrow ZnO(s) + CO_2(g)$.
Which of the following statements about the action of heat on zinc carbonate is (are) true? On heating in a test tube

(*i*) the mass of the sample decreases,
(*ii*) the solid changes from white to yellow and on cooling it changes back to white,
(*iii*) the number of moles of carbon dioxide produced is the same as the number of moles of zinc carbonate used.

Type 2 (See instructions on page 1.)

19 The element X forms the X^{2-} ion. Which of the following compounds have correct formulae?
(*i*) NaX (*ii*) CaX (*iii*) AlX_3 (*iv*) H_2X

Assertion-reason

(See instructions on pages 1–2.)

20 Iron(II) sulphate and iron(III) sulphate are composed of the same three elements. — An iron(II) ion has one less positive charge than an iron(III) ion.

21 The formula for ammonium sulphate is $(NH_4)_2SO_4$. — The ammonium and sulphate ions are NH_4^+ and SO_4^{2-} respectively.

4 Structure of the atom

Simple multiple choice

1 The nuclei of atoms contain only

A protons.
B neutrons.
C electrons.
D protons and neutrons.
E protons and electrons.

2 A particle contains 20 protons, 20 neutrons and 18 electrons. It is therefore
A an atom. C a negative ion. E a molecule.
B a positive ion. D an anion.

3 Particles containing the same number of protons and electrons but different numbers of neutrons are called
A atoms. B isomers. C isotopes. D atoms. E allotropes.

4 The number of neutrons in a $^{238}_{92}U$ atom is
A 92. B 146. C 238. D 330.

5 The electron arrangement of an atom containing 15 electrons is
A 2, 13. B 2, 8, 5. C 2, 4, 8, 1. D 2, 2, 8, 3.
E some arrangement other than the above.

6 Isotopes are atoms of the same element containing different numbers of
A protons. C neutrons.
B electrons. D protons and electrons.

7 The change from Fe to Fe^{3+} involves the
A loss of electrons. C loss of neutrons. E gaining of electrons.
B loss of protons. D gaining of protons.

8 The change from Fe^{3+} to Fe^{2+} involves the
A loss of an electron. C loss of a neutron. E gaining of an electron.
B loss of a proton. D gaining of a proton.

9 An atom of aluminium can be represented by $^{27}_{13}Al$. Which one of the following statements is true?
A has a mass number of 13. C has an atomic number of 27.
B contains 13 neutrons. D contains 14 electrons.
E contains one more neutron than electron.

10 When an atom loses an electron it forms
A a noble gas atom. C a negative ion.
B a positive ion. D a metal atom.
E an isotope of the original atom.

11 A particle which contains 35 protons, 45 neutrons and 36 electrons is
A a noble gas atom. C a negative ion. E a halogen atom.
B a positive ion. D a metal atom.

12 When an oxygen atom is converted to an oxide ion the
A oxygen atom loses two protons. C oxygen atom loses two electrons.
B oxygen atom gains two electrons. D atomic number increases by two.
E atomic number decreases by two.

Classification

Questions 13–15. Select the correct answer A–E from the list below.
A 9 B 10 C 18 D 19 E 28
An atom contains 9 protons, 10 neutrons and 9 electrons.

13 How many particles are there in the nucleus of the atom?

14 What is the mass number of the atom? **15** What is the atomic number of the atom?

Questions 16–21 refer to five atoms labelled A to E.

	Mass no.	*Atomic no.*
A	40	20
B	19	9
C	7	3
D	16	8
E	14	7

Which one of these atoms

16 contains 7 protons?

17 is an isotope of the element oxygen?

18 has an electron arrangement 2, 7?

19 contains the greatest number of neutrons?

20 is in Group I of the Periodic Table?

21 contains the least number of electrons?

Multiple completion

Type 1 (See instructions on page 1.)

22 Potassium has an atomic number of 19. Which of the following statements is (are) true for an atom of potassium-39?
Potassium-39 contains
(*i*) 19 protons. (*ii*) 19 electrons. (*iii*) 19 neutrons.

23 All isotopes of hydrogen have atoms containing
(*i*) one proton. (*ii*) one electron. (*iii*) one neutron.

Type 2 (See instructions on page 1.)

24 Atoms of chlorine-35 and chlorine-37
(*i*) contain the same number of protons.
(*ii*) contain the same number of electrons.
(*iii*) contain the same number of neutrons.
(*iv*) are isotopes.

25 Which of the following represent the electron arrangements in noble gases?
(*i*) 2 (*ii*) 2,6 (*iii*) 2,8,8 (*iv*) 2,8,8,2

26 The relative atomic masses of carbon and magnesium are 12 and 24 respectively. Which of the following statements about an atom of carbon is (are) true?
An atom of carbon has
(*i*) a mass of 12 g.
(*ii*) half the mass of an atom of magnesium.
(*iii*) twelve times the mass of a molecule of hydrogen.
(*iv*) twelve times the mass of an atom of hydrogen.

Assertion-reason

(See instructions on pages 1–2.)

27 The numbers of protons and neutrons in any atom are equal. — Protons and neutrons are approximately equal in mass.

28 The nucleus of an atom is positively charged. — The nucleus is composed of positive and neutral particles.

29 All relative atomic masses are calculated using H = 1 as standard. — Hydrogen has the lowest relative atomic mass of all the elements.

5 Bonding and structure

Simple multiple choice

1 Atoms of a metal
A lose electrons to form positive ions.
B gain electrons to form negative ions.
C lose electrons to form positive ions.
D gain electrons to form positive ions.

2 The electron arrangement in an aluminium **ion** is
A 2,8,3. B 2,8,8. C 2,8. D 2,8,5.

3 The electron arrangements of lithium and oxygen atoms are 2,1 and 2,6 respectively.
Which one of the following is the most likely formula of lithium oxide?
A LiO B Li_2O C LiO_2 D Li_2O_2 E Li_2O_3

4 During the formation of a covalent bond the atoms
A share electrons.
B gain or lose electrons.
C share protons.
D gain or lose protons.
E share neutrons.

5 A compound X is composed of two elements. Which one of the following properties is the **best** indication of whether the bonding is ionic or covalent?
A X is almost insoluble in water.
B X does not conduct electricity when solid.
C X is a crystalline solid.
D X conducts electricity when molten.

6 A compound was found to be a non-conductor of electricity when solid but a conductor when melted. This suggests that when the compound is solid it contains
A a giant structure of ions.
B molecules.
C a giant structure of atoms.
D a random arrangement of atoms.
E metallic bonding.

7 Silicon(IV) oxide (silicon dioxide) is composed of a giant structure of atoms. It therefore
A contains one type of atom.
B melts at a low temperature.
C conducts electricity when molten.
D contains free molecules.
E has covalent bonds between silicon and oxygen atoms.

Classification

Questions 8–12

Table 5.1 shows some of the properties of five substances A–E.

Substances	Melting point (°C)	Boiling point (°C)	Electrical conductivity: When solid	Electrical conductivity: When melted	Effect of heating in air
A	800	1470	None	Good	No reaction
B	98	880	Good	Good	Burns to form a solid oxide
C	−30	58	None	None	Burns to form carbon dioxide and steam
D	114	444	None	None	Burns to form a gaseous oxide
E	1700	2200	None	None	No reaction

Table 5.1

From the substances A–E select the substance which could be

8 an ionic compound. **9** a metal.

10 a liquid at room temperature and pressure.

11 one having a giant molecular structure.

12 a compound composed of relatively small molecules.

Questions 13–16

A ionic bonding
B covalent bonding
C co-ordinate bonding
D hydrogen bonding
E metallic bonding

Which of the types of bonding A–E above is present

13 in a piece of copper?

14 within a water molecule?

15 in sodium chloride?

16 between water molecules in liquid water?

Multiple completion

Type 1 (See instructions on page 1.)

17 Which of the following elements form chlorides containing ionic bonding?
(*i*) sodium (*ii*) carbon (*iii*) helium

18 Which of the following statements about magnesium oxide is (are) correct?
Magnesium oxide
(*i*) contains Mg^{2+} and O^{2-} ions.
(*ii*) is a refractory.
(*iii*) contains ionic bonding.

19 Which of the following substances is (are) composed of molecules containing covalent bonds?
(*i*) dry hydrogen chloride (*ii*) magnesium chloride (*iii*) helium

20 Which of the following are allotropes (polymorphs)?
(*i*) liquid oxygen and gaseous oxygen
(*ii*) sulphur dioxide and sulphur(VI) oxide (sulphur trioxide)
(*iii*) white phosphorus and red phosphorus

21 Barium fluoride (BaF_2) forms an ionic lattice. In this lattice
(*i*) there are equal numbers of barium and fluoride ions.
(*ii*) the ions are held together by strong electrostatic forces.
(*iii*) the ions are regularly arranged.

Type 2 (See instructions on page 1.)

22 Hydrogen fluoride (HF) contains hydrogen bonding because of
(*i*) its high boiling point. (*ii*) its high acidity.
(*iii*) the high reactivity of fluorine.
(*iv*) the electronegativity difference between fluorine and hydrogen.

23 Which of the following contain hydrogen bonding?
(*i*) hydrogen, H_2 (*ii*) ethanol, C_2H_5OH (*iii*) methane, CH_4 (*iv*) water, H_2O

Assertion-reason

(See instructions on pages 1–2.)

24 Many compounds containing covalent bonds are liquid at room temperature and pressure. — The forces between molecules containing covalent bonding are usually weak.

Structured questions

25 Three elements P, Q and R have atomic numbers between 2 and 10. An atom of P has one electron less than a noble gas. An atom of Q has three electrons less than a noble gas and R is an alkali metal (group I in the Periodic Table).

(*a*) Write the simplest formula for the compound formed between
(*i*) P and R. (*ii*) P and Q. (*iii*) Q and R.

(*b*) Explain the bonding found
(*i*) between two atoms of P. (*ii*) in the compound of P and R.
(*iii*) in the compound of P and Q.

(*c*) Which of the elements P, Q and R would
(*i*) be non metals? (*ii*) conduct electricity at room temperature and pressure?

6 States of matter

Simple multiple choice

1 In which of the following are the particles most disordered?
A ice at 0°C B water at 0°C C water at 100°C D water vapour at 100°C

2 When steam condenses the particles
A gain energy.
B move further apart.
C vibrate much faster.
D move closer together.
E remain completely unchanged.

3 Which one of the following gives the best evidence that matter consists of tiny moving particles?
A Air can be compressed in a gas syringe by increasing the pressure.
B Many elements conduct electricity.
C A thin layer is formed when a drop of oil is placed on the surface of water.
D When a bottle of perfume is opened the smell is quickly detected in all parts of the room.
E A small mass of water produces a much larger volume of steam.

4 When a few drops of ethanol are placed on the back of the hand, the hand feels cold. This is because ethanol
A requires heat to react.
B freezes on the hand.
C extracts heat to assist evaporation.
D reacts with the surface of the skin.

5 The spontaneous movement of particles of a gas to fill all of the available space is called
A evaporation. B condensation. C diffusion. D Brownian motion.

6 The volume of one mole of water is approximately 18 cm^3. When one mole of water is converted to steam, it has a volume of approximately 30 000 cm^3. This difference in volume is due to the fact that
A steam molecules are larger than water molecules.
B steam contains more molecules than water.
C the mass of steam is greater than the mass of water.
D the space between the molecules is greater in steam than in water.

7 The motion of fine pollen grains on the surface of water (Brownian movement) is caused by
A collisions between pollen grains and water molecules.
B convection currents in the water.
C attraction and repulsion between charged particles.
D diffusion of pollen grains.
E chemical reaction between pollen grains and the water.

8 Increasing the pressure on a liquid
A decreases its volume.
B makes the particles of the liquid move much closer together.
C makes it solidify. D increases its mass. E increases its boiling point.

9 When smoke particles are illuminated and viewed through a microscope, the particles are observed to move with a jerky, random motion. Which one of the following best explains this observation?
A Electrostatic charges are caused by particles rubbing together.
B Light energy excites the particles making them move faster.

C The particles are charged and repel each other.
D The smoke particles collide with invisible particles in the air.

10 A gas jar containing dark brown nitrogen dioxide gas is placed upside down over a gas jar containing air. Which one of the following best describes the colours inside the gas jars after a long period of time? (Nitrogen dioxide has a higher density than air.)

	Upper gas jar	*Lower gas jar*
A	dark brown	colourless
B	light brown	light brown
C	dark brown	light brown
D	colourless	dark brown
E	light brown	dark brown

11 The braking system of a car contains a liquid. The reason for using a liquid is because it
A cannot be compressed.
B readily turns to a gas.
C takes up the shape of the container.
D does not have a fixed shape.
E boils at a constant temperature.

Questions 12 and 13 refer to Fig. 6.1. The figure shows a cooling curve for a liquid organic acid which is cooled by immersing the test tube containing the acid in a bath of water.

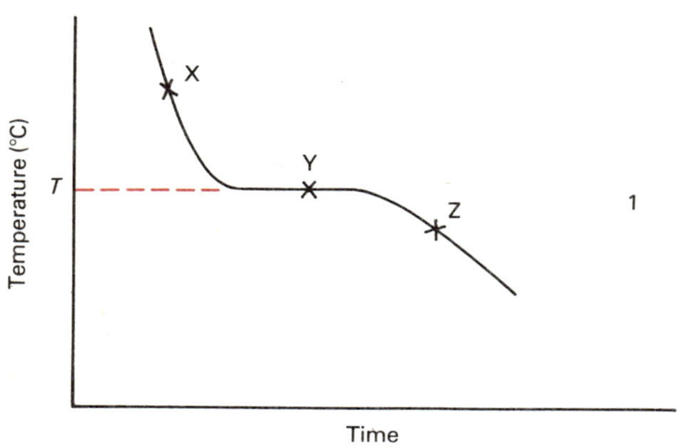

Fig. 6.1

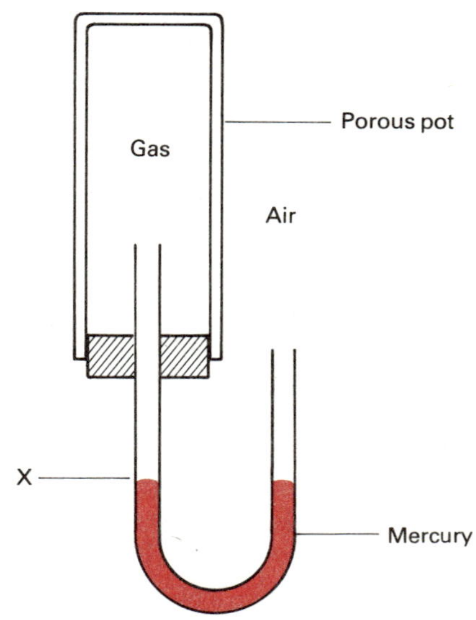

Fig. 6.2

12 Which one of the following gives the correct states of the organic acid at X, Y and Z?

	X	Y	Z
A	gas	liquid	solid
B	liquid	solid	solid
C	liquid	solid and liquid	solid
D	solid	liquid	gas
E	liquid	liquid	solid and liquid

13 The temperature of T °C represents
A the boiling point of the organic acid.
B the melting point of the organic acid.
C room temperature.
D the temperature of the cooling water.

14 50 cm^3 of carbon monoxide (CO) diffused in 25 seconds. How long will it take for the same volume of ethene (C_2H_4) to diffuse under the same conditions of temperature and pressure? ($A_r(H)=1$, $A_r(C)=12$, $A_r(O)=16$)
A 15 seconds
B 25 seconds
C 30 seconds
D 35 seconds
E 45 seconds

15 The apparatus shown in Fig. 6.2 was left standing in air. Which one of the following gases when placed in the porous pot would cause the mercury level at X to rise?
A air
B carbon dioxide
C sulphur dioxide
D helium
E nitrogen dioxide

Classification

Questions 16–20

A melting B condensing C boiling D freezing E subliming

Which of the above processes best describe each of the following changes?

16 Liquid candle wax turns to a waxy solid on cooling.

17 Ice cubes turn to water at room temperature.

18 Ethanol turns to ethanol vapour at 78°C.

19 Ammonium chloride turns to a vapour on heating and reforms to the solid on cooling.

20 Propanone vapour turns to liquid propanone on cooling.

Questions 21–25 refer to the cooling curve in Fig. 6.3.

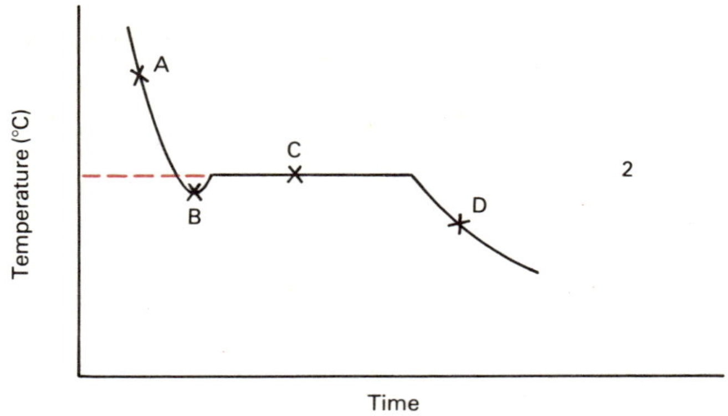

Fig. 6.3

Which letter represents the part of the graph where

21 the liquid is losing heat without lowering the temperature.

22 supercooling is taking place.

23 the rate of temperature change is greatest.

24 both liquid and solid exist together.

25 only solid is present.

Multiple completion

Type 1 (See instructions on page 1.)

26 The particles in a gas are
(*i*) regularly arranged. (*ii*) tightly packed together. (*iii*) in random motion.

27 The temperature at which pure ethanol boils
(*i*) gradually increases as the ethanol boils away.
(*ii*) depends upon the temperature of the room.
(*iii*) falls slightly if atmospheric pressure falls.

28 A crystal of potassium manganate(VII) is dropped into a stoppered measuring cylinder full of water. After a few hours, the solution is a pale pink colour throughout.
Which of the following processes have occurred?
(*i*) diffusion (*ii*) dissolving (*iii*) evaporation

Type 2 (See instructions on page 1.)

29 Which of the following gases diffuse faster than nitrogen (relative molecular mass $M_r = 28$)?
(*i*) Hydrogen ($M_r = 2$)
(*ii*) Methane ($M_r = 16$)
(*iii*) Ammonia ($M_r = 17$)
(*iv*) Oxygen ($M_r = 32$)

30 A liquid
(*i*) takes up the shape of the container.
(*ii*) has a definite shape.
(*iii*) expands slightly on heating.
(*iv*) contains a regular arrangement of particles.

Assertion-reason

(See instructions on pages 1–2.)

31	The volume of 1 mole of water molecules is less than 1 mole of ice at 0°C.	Some water molecules are lost when ice melts.
32	The change from liquid methanol to methanol vapour is a change of state.	The methanol molecules change in composition during the change from vapour to liquid.

33 When water vapour is cooled it turns to a liquid. — When water vapour is cooled the kinetic energy of the molecules decreases.

34 The particles of a liquid at room temperature are in a state of random motion. — The movement of the particles in a liquid is determined by collisions between molecules.

35 When a liquid evaporates the particles possessing most kinetic energy escape from the liquid. — Evaporation is accompanied by a cooling of the liquid.

36 The random movement of pollen grains on the surface of water is called Brownian motion. — This movement is caused only by collisions between pollen grains.

Structured questions

37 When ammonia and hydrogen chloride gases mix, dense white fumes are formed.

$NH_3(g) + HCl(g) \rightarrow NH_4Cl(s)$

A long, dry glass tube was clamped horizontally. A piece of cotton wool, soaked in concentrated ammonia solution, and a piece of cotton wool, soaked in concentrated hydrochloric acid, were placed in opposite ends of the glass tube at the same time. After ten minutes a white solid ring was formed as shown in Fig. 6.4.

Fig. 6.4

(*a*) Name the white solid formed in the tube.
(*b*) Explain why the white ring was formed closer to the end containing the pad of hydrochloric acid.
(*c*) The molecules of ammonia and hydrogen chloride are moving rapidly at room temperature. Give two reasons why the formation of the ring of white solid takes ten minutes.
(*d*) What name is given to this movement of molecules?

A similar experiment can be carried out using saturated solutions of hydrogen sulphide (H_2S) and sulphur dioxide (SO_2) in place of ammonia solution and hydrochloric acid.

$2H_2S(g) + SO_2(g) \rightarrow 3S(s) + 2H_2O(l)$

(*e*) What is the solid formed in this case?
(*f*) Where is the solid formed in the tube?
($A_r(H) = 1$, $A_r(O) = 16$, $A_r(S) = 32$)*

7 Some laws relating to gases

Simple multiple choice

1 A gas at s.t.p. is at
A room temperature and 760 mm pressure.
B 0°C and 760 mm pressure.
C 0 K and 760 mm pressure.
D room temperature and pressure.

2 Which one of the following mathematical expressions is the best summary of Boyle's law?
A $P \times V$ is a constant.
B $P \times V$ is a constant at constant temperature.
C $P \times V$ is a constant at constant pressure.
D V/T is a constant.
E V/T is a constant at constant pressure.

3 If the volume of a fixed mass of gas is kept constant and the temperature is increased
A the pressure increases.
B the molecules move more slowly.
C the pressure decreases.
D the spaces between the molecules increase.
E the number of collisions between molecules decreases.

* [$A_r(O) = 16$ or $O = 16$. This gives the relative atomic mass of oxygen as 16. In a similar way the relative molecular mass of water may be shown as $M_r(H_2O) = 18$.]

4 A fixed mass of gas occupies 2000 cm^3 at a temperature of 30°C and one atmosphere pressure. What would be the new volume of the gas if the temperature was increased to 60°C and the pressure decreased to 0.5 atmosphere?
A 500 cm^3 B 2000 cm^3
C 4000 cm^3 D Between 4000 and 5000 cm^3 E Over 6000 cm^3

5 A fixed mass of argon gas has a pressure of one atmosphere at 20°C. Assuming it is kept in a container of fixed volume, what will be the pressure of the gas at 60°C?

A ⅓ atmospheres C 3 atmospheres E $\frac{293}{333}$ atmospheres

B 1 atmosphere D $\frac{333}{293}$ atmospheres

Classification

Questions 6–9

A volume exactly doubles
B volume increases
C volume remains unchanged
D volume decreases
E volume exactly halves

Which of the above best describes the change in volume which would occur when the following change in conditions takes place? Initially the fixed mass of gas is at 20°C and one atmosphere pressure.

6 The temperature is increased to 30°C and the pressure is unchanged.

7 The temperature is unchanged and the pressure is decreased to 0.5 atmospheres.

8 The gas is transferred to a vessel of twice the volume.

9 Half of the mass of gas escapes but the temperature and pressure are unchanged.

Questions 10–13

A 250 B 500 C 1000 D 2000 E 2500

Select the correct answer A–E from the list above to complete the following table. The table refers to a fixed mass of gas.

Question number	*Volume of gas (cm^3)*	*Pressure (atmospheres)*	*Temperature (K)*
	1000	1	500
10		1	1000
11		2	500
12		2	1000
13	1000	0.5	

Multiple completion

Type 1 (See instructions on page 1.)

14 At 20°C and one atmosphere pressure a sample of gas occupies 1000 cm^3. Under which of the following sets of conditions will the volume be 2000 cm^3?
(*i*) 30°C and 2 atmospheres pressure (*ii*) 10°C and 1 atmosphere pressure
(*iii*) 20°C and 0.5 atmosphere pressure

8 Air and combustion

Simple multiple choice

1 The most common element in air is
A oxygen. B nitrogen. C carbon dioxide. D argon. E hydrogen.

2 Oxygen can be removed from the air by passing the air over
A heated copper(II) oxide.
B heated copper turnings.
C anhydrous calcium chloride.
D sodium hydroxide pellets.

3 Which one of the following substances is produced when a dry sample of rust is heated in air?
A water B oxygen C iron D sulphur dioxide

4 When air is liquefied and allowed to boil, nitrogen boils first and then oxygen. This is because
A nitrogen has a higher boiling point than oxygen.
B oxygen is less soluble than nitrogen.
C nitrogen is less soluble than oxygen.
D oxygen has a higher boiling point than nitrogen.

5 Which one of the following gases is not normally present in air?
A water vapour B argon C krypton D hydrogen E nitrogen

6 A candle was burnt in a limited volume of air until the candle was extinguished. How much oxygen remained in the air at the end of the experiment?
A none B between 0 and 20% C between 20 and 30% D over 30%

Questions 7 and 8

A sample of water weed was placed in water and exposed to sunlight. Bubbles of gas were seen to form on the surface of the leaves.

7 The gas forming was
A oxygen. B hydrogen. C carbon dioxide. D air. E nitrogen.

8 The process taking place was
A respiration. C decomposition. E oxidation.
B evaporation. D photosynthesis.

9 Which one of the following processes decreases the amount of carbon dioxide in the air?
A respiration C decaying of vegetation
B combustion of hydrocarbons D photosynthesis

10 Which one of the following equations correctly represents the burning of zinc in oxygen?
A $Zn(s) + O(g) \rightarrow ZnO(s)$ C $2Zn(s) + O_2(g) \rightarrow 2ZnO(s)$
B $Zn(s) + O_2(g) \rightarrow ZnO_2(s)$ D $2Zn(s) + 2O_2(g) \rightarrow 2ZnO_2(s)$

Questions 11–14

The apparatus in Fig. 8.1 is commonly used to determine the percentage of oxygen in air.

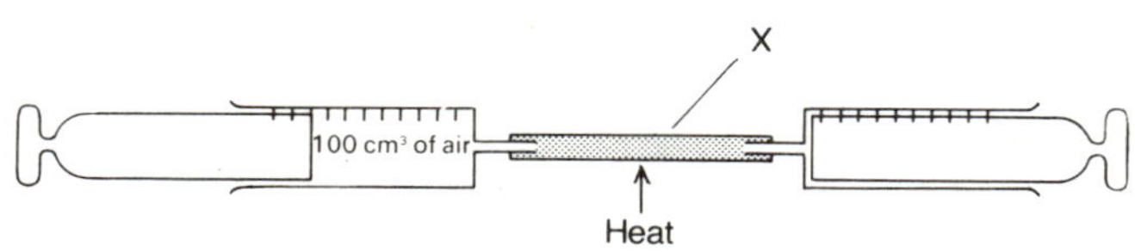

Fig. 8.1

11 Which one of the following substances could be placed at X in order to carry out the experiment safely?
A phosphorus C carbon E iron wire
B copper(II) oxide D sodium hydroxide

12 100 cm^3 of dry air was placed in one of the syringes and passed to and fro across the substance at X while it was being heated. What volume of gas, in cm^3, remained at the end of the experiment?
A 0 B 20 C 60 D 80 E 100

13 What mass of product would have to be formed in this experiment if 1.0 g of X had been used?
A none B 0.25 g C 0.90 g D 1.10 g E 1.80 g

14 The gas which remained at the end of the experiment was
A pure nitrogen. B a mixture of nitrogen and hydrogen.
C pure hydrogen. D pure carbon dioxide.
E a mixture of nitrogen, carbon dioxide and noble gases.

Classification

Questions 15–17

A water vapour B carbon dioxide C carbon monoxide D oxygen

Use the alternatives above to answer the following questions.

A stream of moist oxygen was passed over heated carbon. The resulting gases were passed through an aqueous solution of potassium hydroxide and then over heated copper. The resulting gas was then collected over water.

Which gas

15 was removed by the aqueous solution of potassium hydroxide?

16 was removed by the heated copper? **17** was collected over water?

Multiple completion

Type 1 (See instructions on page 1.)

18 During photosynthesis a plant
(*i*) absorbs carbon dioxide. (*ii*) evolves oxygen. (*iii*) absorbs light energy.

19 Which of the following take place during respiration?
(*i*) oxygen is used up. (*ii*) carbon dioxide is produced. (*iii*) energy is taken in.

20 Which of the following processes use up oxygen?
(*i*) combustion of ethanol (*ii*) respiration (*iii*) photosynthesis

21 Rusting can be prevented by excluding
(*i*) nitrogen. (*ii*) oxygen. (*iii*) water vapour.

Type 2 (See instructions on page 1.)

22 Which of the following would decrease the percentage of oxygen in an enclosed volume of air?
(*i*) rusting of iron
(*ii*) burning a candle
(*iii*) decomposing hydrogen peroxide
(*iv*) respiration

23 Which of the following statements about air is (are) true?
(*i*) Air is a mixture of gaseous elements. (*ii*) The composition of air is fixed.
(*iii*) Liquid air cannot be split up by fractional distillation.
(*iv*) Air contains an active gas.

Assertion-reason

(See instructions on pages 1–2.)

24 Substances burn better in oxygen than in air.	Air is largely nitrogen and nitrogen is an inactive gas.
25 Iron rusts in contact with dry air.	Air contains oxygen which is an active gas.

9 Oxygen and oxides

Simple multiple choice

1 Which one of the following substances will produce an acidic solution when added to water?
A nitrogen B sodium C calcium oxide D sulphur dioxide E copper(II) oxide

2 The catalyst used in the decomposition of hydrogen peroxide to produce oxygen is
A magnesium. B manganese. C magnesium oxide.
D manganese(II) oxide. E manganese(IV) oxide.

3 Which one of the following equations correctly represents the decomposition of hydrogen peroxide?
A $2H_2O_2(aq) \rightarrow 2H_2O(g) + O_2(g)$
B $H_2O_2(aq) \rightarrow H_2O(l) + O(g)$
C $2H_2O_2(aq) \rightarrow 2H_2O(l) + O_2(g)$
D $H_2O_2(aq) \rightarrow H_2(g) + O_2(g)$
E $2H_2O_2(aq) \rightarrow 2H_2O(l) + 2O(g)$

4 Which one of the following methods could be used to produce some oxygen gas?
A reaction of calcium with cold water
B heating potassium manganate(VII) (potassium permanganate) C filtering dry air
D electrolysis of a concentrated solution of sodium chloride

5 Which one of the substances in bold type is a neutral oxide?
A a **gas** which dissolves in water to form a solution with a pH of 5
B a **gas** which dissolves in water to form a solution which has no effect on Universal indicator
C a **solid** which is insoluble in water but is soluble in boiling sodium hydroxide solution
D a **solid** which reacts with both acid and alkali forming salts

6 Which one of the following is an amphoteric oxide?
A carbon dioxide
B carbon monoxide
C zinc oxide
D calcium oxide
E sulphur dioxide

7 Which one of the elements in bold type could **not** be a metal?
A a **dark grey solid** which burns in oxygen to form an oxide of pH 5
B a **grey solid** which conducts heat and electricity
C a **grey solid** which does not react with oxygen
D a **silvery liquid** which burns in oxygen to form an oxide of pH 7
E a **silvery solid** which reacts with dilute hydrochloric acid to produce hydrogen

8 Which one of the following substances burns in excess oxygen to form a single oxide?
A hydrogen sulphide B carbon monoxide C methane D ammonia E ethene

Classification

Questions 9–13
A an acidic oxide which is soluble in water B an acidic oxide which is insoluble in water
C a neutral oxide D a basic oxide which is insoluble in water
E a basic oxide which is soluble in water

From the above list select the group A–E to which each of the oxides in questions **9–13** belong.

9 carbon monoxide 10 carbon dioxide 11 copper(II) oxide

12 silicon(IV) oxide (silicon dioxide) 13 calcium oxide

Multiple completion

Type 1 (See instructions on page 1.)

14 Oxygen is produced by
(*i*) catalytic decomposition of hydrogen peroxide.
(*ii*) action of heat on potassium nitrate.
(*iii*) action of heat on silver(I) oxide.

15 A metal oxide which reacts with molten sodium hydroxide could be
(*i*) basic. (*ii*) acidic. (*iii*) amphoteric.

16 Which of the following elements will burn in air to form a solid oxide?
(*i*) carbon (*ii*) sulphur (*iii*) iron

Type 2 (See instructions on page 1.)

17 Which of the following oxides produce solutions with a pH less than 7 when mixed with water?
(*i*) carbon dioxide (*ii*) magnesium oxide (*iii*) sulphur dioxide (*iv*) carbon monoxide

18 Which of the following compounds decompose on heating to form oxygen as the only gaseous product?
(*i*) red lead (*ii*) potassium nitrite
(*iii*) potassium manganate(VII) (potassium permanganate) (*iv*) lead(II) nitrate

Assertion-reason

(See instructions on pages 1–2.)

	Assertion	Reason
19	All nitrogen oxides are acidic oxides.	Non metals form only acidic oxides.
20	Sodium, magnesium and calcium all form basic oxides which react with water to form alkalis.	Alkalis are formed by metals high in the reactivity series.

10 Hydrogen

Simple multiple choice

1 Two gases, hydrogen and oxygen, are mixed together and exploded. Which one of the following substances is the main product from the reaction that occurs?
A pure hydrogen B liquid oxygen C water D salt water

2 In which one of the following is hydrogen **not** produced?
A reaction of aluminium with aqueous sodium hydroxide solution
B reaction of copper with dilute hydrochloric acid
C electrolysis of an aqueous solution of sodium chloride
D reaction of heated iron with steam
E reaction of sodium hydride with cold water

3 Which one of the following pairs can be used to prepare hydrogen?
A magnesium and dilute hydrochloric acid
B copper and dilute sulphuric acid C copper and steam
D copper and dilute nitric acid E silver and dilute hydrochloric acid

4 It is possible to collect hydrogen by upward delivery (i.e. in an upside down gas jar) because hydrogen
A is insoluble in water. C burns with a squeaky pop. E is colourless.
B is much less dense than air. D mixes readily with air.

5 Which one of the following equations represents the ionic equation for the reaction of zinc with a dilute solution of hydrochloric acid?
A $Zn(s) + H^+(aq) \rightarrow Zn^+(aq) + H(g)$
B $Zn(s) + H^+(aq) \rightarrow Zn^{2+}(aq) + H^-(aq)$
C $Zn(s) + 2H^+(aq) \rightarrow Zn^{2+}(aq) + H_2(g)$
D $Zn(s) + 2H^+(aq) \rightarrow Zn^{2+}(aq) + 2H(g)$
E $Zn(s) + 2HCl(aq) \rightarrow Zn^{2+}(aq) + 2Cl^-(aq) + H_2(g)$

Questions 6 and 7
For these questions note the following order of reactivity of some common metals:
Potassium (most reactive), calcium, magnesium, zinc, iron, lead, copper, silver (least reactive).

6 Which one of the following metal oxides is unchanged when heated in a stream of hydrogen?
A iron(III) oxide C lead(II) oxide E calcium oxide
B silver(I) oxide D copper(II) oxide

7 Which one of the following metals does **not** produce hydrogen from water or steam?
A magnesium B potassium C copper D calcium E zinc

Multiple completion

Type 1 (See instructions on page 1.)

8 In which of the following reactions is hydrogen produced?
(*i*) adding iron to dilute hydrochloric acid and warming
(*ii*) heating solid sodium hydrogencarbonate
(*iii*) adding copper to dilute nitric acid and warming

9 Which of the following metal oxides is (are) reduced by heating in a stream of hydrogen?
(*i*) copper(II) oxide (*ii*) lead(II) oxide (*iii*) magnesium oxide

Assertion-reason

(See instructions on pages 1–2.)

10 Burning a jet of hydrogen gas in the laboratory can be extremely dangerous. | Mixtures of hydrogen and air are explosive.

Structured questions

11 There are three isotopes of the element hydrogen – hydrogen, deuterium and tritium. They have slightly different physical properties but identical chemical properties. The properties of the three isotopes are summarized in Table 10.1.

Isotope	Symbol	Structure	% Occurrence
Hydrogen	H	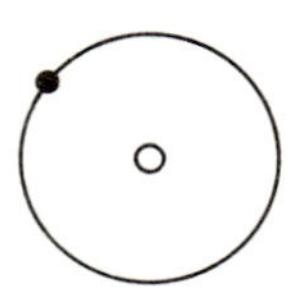	99.985
Deuterium	D		0.015
Tritium	T		traces

Key: ● is an electron
⬤ is a neutron
○ is a proton

Table 10.1

(*a*) Define the term 'isotope'.
(*b*) What are the mass numbers of the three isotopes?
(*c*) Why is the relative atomic mass of hydrogen 1.007 97 rather than exactly 1?
(*d*) Draw the structures of H^+, D^- and D_2.
(*e*) Why is the hydrogen gas that is evolved in chemical reactions given the formula H_2 and not D_2 or T_2?

Heavy water D_2O is water containing deuterium instead of ordinary hydrogen atoms. Heavy water can be obtained in two ways:

I *Repeated electrolysis of water.* During the electrolysis of water, hydrogen is produced at a rate six times faster than deuterium. After electrolysis the remaining water is much richer in deuterium. Heavy water can thus be obtained by continued electrolysis of water.

II *Repeated fractional distillation.* There is a slight difference in boiling point between H_2O and D_2O. After repeated fractional distillation, heavy water remains in the distillation flask.

The chemical properties of heavy water and ordinary water are identical.

(*f*) Why is heavy water extremely expensive to produce?
(*g*) Give a physical property which is different for water and heavy water.
(*h*) Explain how a sample of deuterium can be obtained from a sample of heavy water using a metal. Write a balanced equation for the reaction.

(*i*) Describe a chemical test to distinguish deuterium from oxygen.
(*j*) Write balanced equations for the reaction of deuterium with
(*i*) copper(II) oxide. (*ii*) nitrogen. (*iii*) chlorine.

11 Water

Simple multiple choice

1 To test whether a liquid is pure water it is necessary to
A find the pH with Universal indicator. B taste it.
C add anhydrous copper(II) sulphate. D add cobalt(II) chloride paper.
E find its boiling point.

2 All hydrated compounds
A contain water of crystallization. B are soluble in water. C are liquids.
D absorb water from the atmosphere. E lose water to the atmosphere.

3 When salt is dissolved in water the salt is called the
A precipitate. B solute. C solvent. D solution.

4 A solid which loses water of crystallization when left standing in air is said to be
A efflorescent. B deliquescent. C anhydrous. D hygroscopic. E hydrated.

5 70 g of a saturated solution of a salt at 30°C were evaporated to dryness and 20 g of solid remained. The solubility of the salt at 30°C in grams per 100 g of water is
A 20 g B 30 g C 40 g D 50 g E 70 g

6 All solutions containing water
A freeze at exactly 0°C and boil at exactly 100°C.
B are colourless.
C turn cobalt(II) chloride paper from blue to pink.
D evaporate to dryness leaving no residue. E have a pH of 7.

7 Soap is produced by heating olive oil with sodium hydroxide solution. Salt is added to the solution and the soap forms a solid. The salt is used in order to
A neutralize the excess sodium hydroxide. B react with the olive oil to form soap.
C soften the water. D remove the excess olive oil. E precipitate out the soap.

8 Which one of the following changes would be observed if a crystalline substance gave up water of crystallization to the air?
A no visible change B the crystals crumble to a powder
C the crystals become a solution D a moist layer is formed on the surface of the crystals
E the crystals change colour

9 The fur produced in an aluminium kettle is caused by
A the reaction between aluminium and boiling water.
B the reaction between aluminium and dissolved substances in the water.
C the decomposition of the substance causing temporary hardness in water.
D the evaporation of water to dryness.

Classification

Questions 10–14
A 0.00 g B 4.00 g C 5.00 g D 5.50 g E 10.00 g

Select from the list above the mass which is most likely to remain when 5.00 g of each of the following substances are left exposed to the atmosphere for a few days.

10 solid carbon dioxide ('Drikold') **11** an efflorescent compound

12 a hygroscopic compound **13** a very volatile liquid **14** a deliquescent compound

Questions 15–21
A sodium carbonate B calcium carbonate C calcium sulphate
D calcium hydrogencarbonate E calcium hydroxide

From the above list select a compound which

15 is the chief constituent of limestone rocks. **16** dissolves in water to form limewater.

17 causes temporary hardness in water. **18** is the major cause of permanent hardness in water.

19 softens water containing temporary and permanent hardness.

20 is formed when rain water trickles through limestone rocks.

21 when added in the correct quantity to temporary hard water removes the hardness but has no effect on permanent hardness.

Multiple completion

Type 1 (See instructions on page 1.)

22 The following table shows the solubility of potassium chlorate(I) at various temperatures.

Temperature (°C)	10	20	30	40	50
Solubility (g per 100 g water)	5	15	30	50	75

Which of the following mixtures would give a saturated solution?

(*i*) 20 g of potassium chlorate(I) in 50 g of water at 40°C
(*ii*) 10 g of potassium chlorate(I) in 100 g of water at 10°C
(*iii*) 40 g of potassium chlorate(I) in 50 g of water at 50°C

23 Which of the following statements about the reactions of metals with water are true?

(*i*) Potassium reacts with cold water to form hydrogen.
(*ii*) Zinc reacts with steam to form zinc oxide.
(*iii*) All metals react with water to form hydrogen.

24 The water cycle in nature involves
(*i*) evaporation. (*ii*) condensation. (*iii*) filtration.

25 Which of the following processes are used in the purification of water for use as tap water?
(*i*) filtration (*ii*) chlorination (*iii*) distillation

Type 2 (See instructions on page 1.)

26 Anhydrous copper(II) sulphate

(*i*) contains no water of crystallization.
(*ii*) is a compound of three elements.
(*iii*) turns blue on the addition of water. (*iv*) is efflorescent.

27 Detergents are produced from the reaction of

(*i*) strong acids on natural fats and oils.
(*ii*) strong alkalis on natural fats and oils.
(*iii*) strong alkalis on hydrocarbons. (*iv*) strong acids on hydrocarbons.

Assertion-reason

(See instructions on pages 1–2.)

	Assertion	Reason
28	Detergents are good wetting agents.	Detergents lower the surface tension of water.
29	Soapless detergents lather well in hard and soft water.	The acid content of the soapless detergent prevents the formation of scum.
30	Calcium and magnesium ions cause hardness in water.	Calcium and magnesium ions are removed from solution in an ion-exchange column.
31	In experiments involving weighing, copper(II) oxide should always be dried before weighing.	Copper(II) oxide is hygroscopic.
32	A desiccator often contains anhydrous calcium chloride as drying agent.	Anhydrous calcium chloride is deliquescent.

Structured questions

33 Figure 11.1 (page 23) shows the solubilities of two salts X and Y in water over a range of temperature. Use this graph to answer the following questions.

(*a*) How does the solubility of X in water change with temperature?
(*b*) What is the solubility of X in water at 40°C?
(*c*) Which of the salts is more soluble in water at
(*i*) 20°C? (*ii*) 60°C?
(*d*) At which temperature are the solubilities of X and Y the same?
(*e*) Find the temperature at which 24 g of X when added to 50 g of water, at this temperature, form a saturated solution with no undissolved solid.

80 g of X and 40 g of Y were added to 100 g of water at 60°C. Some crystals remained undissolved. These crystals were removed and the solution cooled to 15°C when more crystals were formed.

(*f*) How many grams of X and Y respectively remain undissolved at
(*i*) 60°C? (*ii*) 15°C?

34 An experiment was carried out to find the ions which cause hardness in water.

5 cm^3 of each of the solutions used were placed in a separate test tube and 20 drops of soap solution added to each. The test tubes were shaken for 15 seconds and the height of the lather produced in each tube was measured. The results are shown in the following table.

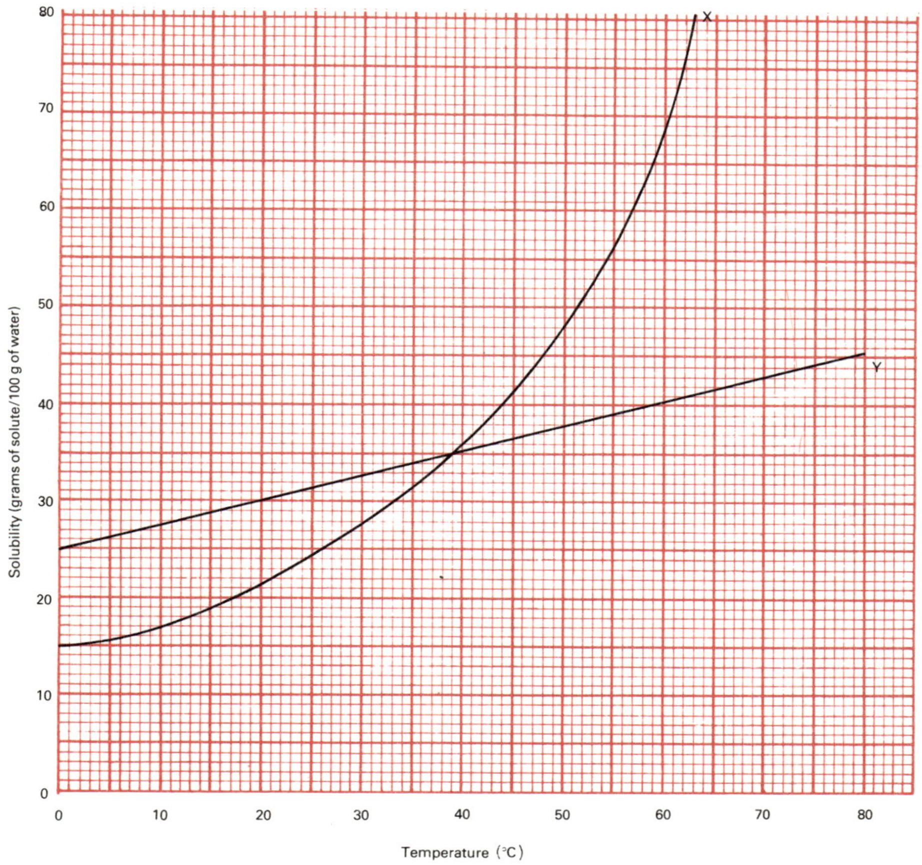

Fig. 11.1

Solution used	*Height of lather (mm)*
Potassium sulphate	20
Magnesium sulphate	4
Sodium sulphate	18
Calcium chloride	3
Potassium chloride	21
Magnesium chloride	3
Sodium chloride	20

(*a*) Name the piece of apparatus most suitable for adding the drops of soap solution.

(*b*) Using the results in the table, name two ions which cause hardness in water.

50 cm^3 of tap water were measured into a conical flask. Soap solution was added in small portions. After each addition the flask was shaken and additions were made until a lasting lather was produced.

The experiment was repeated with

(*i*) 50 cm^3 of boiled tap water. (*ii*) 50 cm^3 of distilled water.

The results are shown in the following table

Water sample	*Volume of soap solution (cm^3)*
tap water	9.5
boiled tap water	0.5
distilled water	0.5

(*c*) Which piece of apparatus is most suitable for

(*i*) measuring out 50 cm^3 of tap water? (*ii*) adding small volumes of soap solution?

(*d*) What type or types of hardness does the tap water contain?

(*e*) Name the chemical or chemicals present in this sample of tap water which cause this hardness.

(*f*) Write a symbol equation for the reaction which causes this hardness to exist in water.

It is possible to buy liquid soaps and liquid soapless detergents.

(*g*) Describe how the results in the above experiment would be different if a liquid soapless detergent had been used instead of a liquid soap solution.

12 The mole and chemical calculations

Simple multiple choice

1 One mole of magnesium atoms contains the same number of atoms as one mole of
A hydrogen, H_2. B oxygen, O_2. C helium, He. D chlorine, Cl_2.

2 The mass of oxygen contained in 0.5 mole of oxygen molecules (O_2) is
A 4 g. B 8 g. C 16 g. D 32 g. E 64 g.
($A_r(O) = 16$)

3 8 g of oxygen contain the same number of atoms as 7 g of
A nitrogen. B helium. C hydrogen. D neon. E chlorine.
($A_r(H) = 1$, $A_r(He) = 4$, $A_r(N) = 14$, $A_r(Ne) = 20$, $A_r(Cl) = 35.5$)

4 What is the mass of one mole of sulphuric acid (H_2SO_4)?
A 58 g B 98 g C 108 g D 198 g E 208 g
($A_r(H) = 1$, $A_r(O) = 16$, $A_r(S) = 32$)

5 What is the mass of one mole of aluminium sulphate $Al_2(SO_4)_3$?
A 1 g B 150 g C 315 g D 327 g E 342 g
($A_r(O) = 16$, $A_r(Al) = 27$, $A_r(S) = 32$)

6 The mass of 0.1 mole of sodium carbonate decahydrate ($Na_2CO_3.10H_2O$) is
A 6.8 g. B 13.3 g. C 14.2 g. D 28.6 g. E 28.7 g.
($A_r(H) = 1$, $A_r(C) = 12$, $A_r(O) = 16$, $A_r(Na) = 23$)

7 A sample of an oxide of antimony (Sb) contained 40 g of antimony combined with 8 g of oxygen. Which one of the following represents the simplest formula for the oxide?
($A_r(Sb) = 120$, $A_r(O) = 16$)
A SbO B SbO_2 C Sb_2O D Sb_2O_3 E Sb_3O_2

8 2.4 g of magnesium are burned in excess oxygen. What mass of magnesium oxide (MgO), in grams, will be formed? ($A_r(O) = 16$, $A_r(Mg) = 24$)
A 1.6 g B 4.0 g C 8.0 g D 20.0 g E 40.0 g

9 20.8 g of lead combine with 3.2 g of oxygen. What is the formula of the oxide produced?
($A_r(O) = 16$, $A_r(Pb) = 208$)
A PbO B PbO_2 C Pb_2O D Pb_2O_3 E Pb_3O_4

10 The percentage, by mass, of iron in Fe_2O_3 is
A below 20%
B between 20 and 40%
C between 40 and 60%
D between 60 and 80%
E above 80%
($A_r(Fe) = 56$, $A_r(O) = 16$)

11 A hydrocarbon contains 80% of carbon and 20% of hydrogen by mass. The empirical (or simplest) formula of the hydrocarbon is
A CH_2. B CH_3. C CH_4. D C_2H_4. E C_2H_6.
($A_r(C) = 12$, $A_r(H) = 1$)

12 The equation for the reduction process taking place in the blast furnace is
$Fe_2O_3(s) + 3CO(g) \rightarrow 2Fe(s) + 3CO_2(g)$
What is the maximum mass of iron which could be extracted from 40 tonnes of iron(III) oxide?
($A_r(O) = 16$, $A_r(Fe) = 56$)
A 8 tonnes B 14 tonnes C 16 tonnes D 28 tonnes E 56 tonnes

13 Which one of the following procedures is most suitable for preparing an exactly 0.1 M solution (0.1 mole per dm^3) of sodium carbonate? ($M_r = 106$)
A dissolving 10.6 g of sodium carbonate and making up to 1000 cm^3 with water
B dissolving 10.6 g of sodium carbonate in 1000 cm^3 of water
C adding 100 cm^3 of M sodium carbonate to 1000 cm^3 of water
D dissolving 10.6 g of sodium carbonate in 100 cm^3 of water

14 A solution of sodium hydroxide in water containing 2.0 g of sodium hydroxide in 250 cm^3 of solution has a concentration of
A 0.02 mole per dm^3.
B 0.1 mole per dm^3.
C 0.2 mole per dm^3.
D 0.25 mole per dm^3.
E 1 mole per dm^3.

15 Which one of the following compounds contains the greatest percentage of nitrogen by mass?
($A_r(H) = 1$, $A_r(N) = 14$, $A_r(O) = 16$, $A_r(Na) = 23$, $A_r(S) = 32$, $A_r(Cl) = 35.5$, $A_r(Ca) = 40$)
A NH_4Cl B $NaNO_3$ C $(NH_4)_2SO_4$ D NH_4NO_3 E $Ca(NO_3)_2$

16 A compound X has an empirical formula of CH_2 and a relative molecular mass of 70 ($A_r(H)=1$, $A_r(C)=12$). The molecular formula of X is
A CH_2. B C_2H_4. C C_3H_6. D C_4H_8. E C_5H_{10}.

Questions 17 and 18

$Mg(s)+2HCl(aq) \rightarrow MgCl_2(aq)+H_2(g)$

One mole of molecules of any gas occupies 24 dm^3 at room temperature and pressure. ($A_r(Mg)=24$, $A_r(Cl)=35.5$)

17 What volume of hydrogen at room temperature and pressure would be produced when 0.1 g of magnesium reacts with excess hydrochloric acid?
A 1.0 cm^3 B 10 cm^3 C 100 cm^3 D 500 cm^3 E 1000 cm^3

18 What mass of anhydrous magnesium chloride would be produced from 2.4 g of magnesium?
A 7.1 g B 9.5 g C 9.7 g D 71 g E 95 g

19 Different volumes of M lead(II) nitrate solution (1 mole per dm^3) were added to six separate portions of M potassium iodide solution (1 mole per dm^3). Each tube was made up to the total volume, centrifuged and the height of the precipitate measured.
The following results were obtained:

Volume of M lead(II) nitrate $(cm)^3$	1	2	3	4	5	6
Height of precipitate (mm)	9	17	22	27	35	35

How many moles of potassium iodide will react with 1 mole of lead(II) nitrate?
A 0.5 B 1 C 1.5 D 2 E 2.5

20 $2KHCO_3(s) \rightarrow K_2CO_3(s)+CO_2(g)+H_2O(l)$
The mass of 1 mole of potassium hydrogencarbonate is 100 g. How many moles of carbon dioxide will be produced if 50 g of potassium hydrogencarbonate are heated?
A 0.25 B 0.5 C 1.0 D 2.0 E 25

21 Which one of the following contains the greatest number of moles of acid?
A 10 cm^3 of 10 M acid
B 25 cm^3 of 5 M acid
C 25 cm^3 of 4 M acid
D 50 cm^3 of 2 M acid
E 100 cm^3 of M acid

Classification

Questions 22–25

	Mass in grams	*Number of moles*
A	1	0.25
B	6	0.5
C	32	1.0
D	2.8	0.05
E	6.4	0.1

The above table contains the masses of five elements labelled A–E and the corresponding number of moles of atoms in each sample. The letters A–E do not represent the usual symbols for the elements.
Select the element A–E which

22 contains 6×10^{23} particles (Avogadro's number).

23 has a relative atomic mass of 12.

24 contains the smallest particles.

25 contains the smallest number of particles.

Multiple completion

Type 1 (See instructions on page 1.)

26 One mole of hydrogen molecules (H_2) at room temperature and pressure
(*i*) has the same mass as one mole of oxygen molecules (O_2).
(*ii*) has the same volume as one mole of oxygen molecules at room temperature and pressure.
(*iii*) contains the same number of molecules as one mole of oxygen molecules.

Type 2 (See instructions on page 1.)

27 One mole of helium atoms and one mole of iron atoms have the same
(*i*) mass. (*ii*) density. (*iii*) volume. (*iv*) number of atoms.

Assertion-reason

(See instructions on pages 1–2.)

28 One mole of chlorine molecules contains the same number of atoms as one mole of fluorine molecules.	Chlorine and fluorine are in the same group of the Periodic Table.

Structured questions

29 Copper(II) oxide is reduced by heating in a stream of hydrogen.

$CuO(s) + H_2(g) \rightarrow Cu(s) + H_2O(g)$

Dry copper(II) oxide was placed inside a combustion tube. The supply of hydrogen was turned on and hydrogen passed through the tube. After waiting some time the surplus hydrogen was burned at a jet. The tube was heated strongly until all the copper(II) oxide had been converted to copper. The apparatus was allowed to cool with hydrogen still passing before dismantling.

(*a*) What colour change accompanied the change from copper(II) oxide to copper?
(*b*) Why was the hydrogen passed through the apparatus for some time before lighting?
(*c*) Why was the apparatus allowed to cool before dismantling?

A series of experiments were carried out to find the formula of an oxide of copper. The results are shown in the following table.

Group	*Mass of copper oxide* (g)	*Mass of copper produced* (g)	*Mass of oxygen produced* (g)
1.	0.62	0.55	0.07
2.	0.90	0.80	0.10
3.	1.12	1.00	0.12
4.	1.69	1.50	
5.	1.80	1.60	

(*d*) Complete the missing values in the table.
(*e*) Plot the mass of copper (on the *y* axis) against the mass of oxygen (on the *x* axis). Draw the best straight line through the points and through the origin.
(*f*) Use your graph to find the mass of oxygen which would combine with 1.28 g of copper.
(*g*) Calculate the mass of oxygen which would combine with 128 g of copper.
(*h*) How many moles of copper atoms does 128 g of copper represent? ($A_r(Cu) = 64$)
(*i*) How many moles of oxygen atoms combine with 128 g of copper? ($A_r(O) = 16$)
(*j*) What is the simplest formula of this oxide of copper?

13 Laws of chemical combination

Simple multiple choice

1 Gay Lussac's law is concerned with the
A number of particles in a fixed mass of gas.
B volumes of gases reacting.
C volume of a gas at different temperatures.
D volume of a gas at different pressures.
E rate of diffusion of gases.

2 If 100 cm^3 of oxygen at room temperature and pressure contain *n* molecules, how many molecules are there in 100 cm^3 of gaseous sulphur dioxide at room temperature and pressure?

A $0.1n$ B $0.5n$ C n D $2n$ E $\frac{5n}{2}$

3 Three samples of zinc chloride were prepared by three different methods. In each case the percentage of zinc, by mass, in each sample is the same. This is in accordance with
A the law of conservation of mass.
B the law of constant composition.
C the law of multiple proportions.
D Avogadro's law.
E Gay Lussac's law.

Questions 4 and 5

$ZnCO_3(s) + 2HCl(aq) \rightarrow ZnCl_2(aq) + H_2O(l) + CO_2(g)$
$BaCl_2(aq) + H_2SO_4(aq) \rightarrow BaSO_4(s) + 2HCl(aq)$
$Pb(NO_3)_2(aq) + 2HCl(aq) \rightarrow PbCl_2(s) + 2HNO_3(aq)$
$FeS(s) + 2HCl(aq) \rightarrow FeCl_2(aq) + H_2S(g)$

4 The above reactions were carried out in an enclosed vessel. In how many of the reactions will there be a mass change?
A 0 B 1 C 2 D 3 E 4

5 The above reactions were repeated in an open beaker. In how many reactions will the beaker and contents change in mass?
A 0 B 1 C 2 D 3 E 4

Questions 6 and 7
The equation for the complete combustion of propane in excess oxygen is

$C_3H_8(g) + 5O_2(g) \rightarrow 3CO_2(g) + 4H_2O(l)$

6 The volume of oxygen required at room temperature and pressure for the complete combustion of 10 dm^3 of propane is
A 10 dm^3. B 20 dm^3. C 30 dm^3. D 40 dm^3. E 50 dm^3.

7 Assuming no unreacted oxygen remains, the total volume of products at room temperature and pressure after the complete combustion of 10 dm^3 of propane is
A 30 dm^3. B 40 dm^3. C 50 dm^3. D 60 dm^3. E 70 dm^3.

8 The equation for the burning of hydrogen sulphide in excess oxygen is
$2H_2S(g) + 3O_2(g) \rightarrow 2H_2O(g) + 2SO_2(g)$.
What volume of sulphur dioxide is produced from the complete combustion of 10 dm^3 of hydrogen sulphide? (All volumes are measured at the same temperature and pressure.)
A 5 dm^3 B 10 dm^3 C 15 dm^3 D 20 dm^3 E 25 dm^3

Assertion-reason

(See instructions on pages 1–2.)

9 The law of multiple proportions is concerned with two elements forming more than one compound. — Each compound contains two elements combined together in fixed proportions irrespective of the method of formation.

10 The decomposition of ammonia produces three times the volume of hydrogen compared to nitrogen. — The formula of ammonia is NH_3.

14 Metals(I)

Simple multiple choice

1 The coating of aluminium with a thin, coloured, decorative coating by means of electrolysis is called
A electroplating. B aluminizing. C anodizing. D chromatography. E dyeing.

2 When metals are annealed they are
A hardened. B melted. C softened. D protected. E burnt.

3 Metals are good electrical conductors because
A the particles are closely packed.
B the arrangement of particles is regular.
C they are shiny in appearance.
D they contain free electrons.

4 Which one of the following properties is typical of those of a non-metal?
A good conductor of heat
B malleable
C forms an acidic oxide
D good conductor of electricity
E high melting point

5 The slow reaction of aluminium with dilute hydrochloric acid is due to
A an unreactive oxide coating on the aluminium.
B the low solubility of aluminium chloride which prevents reaction.
C the low position of aluminium in the reactivity series.
D the acid not being concentrated.

6 Which one of the following contains the highest percentage of iron?
A tool steel
B stainless steel
C pig iron
D iron ore
E wrought iron

7 Which one of the following alloys does not contain copper?
A bronze B brass C solder D duralumin

Classification

Questions 8–13

A calcium
B calcium oxide
C calcium carbonate
D calcium hydroxide
E calcium hydrogencarbonate

From A–E above, select the best answer to questions **8–13** (page 28). The framework diagram in Fig. 14.1 (page 28) may help in answering these questions.

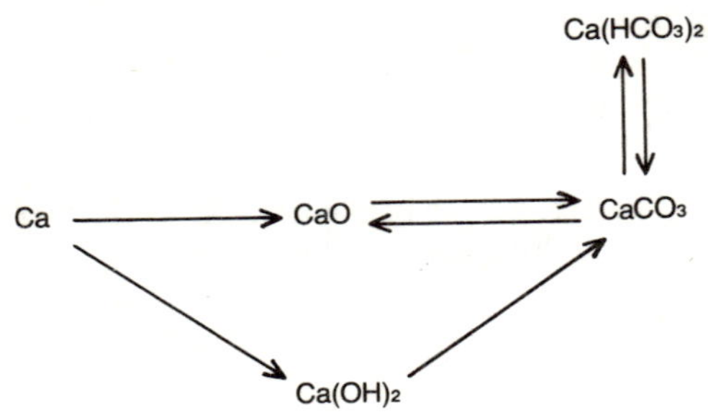

Fig. 14.1

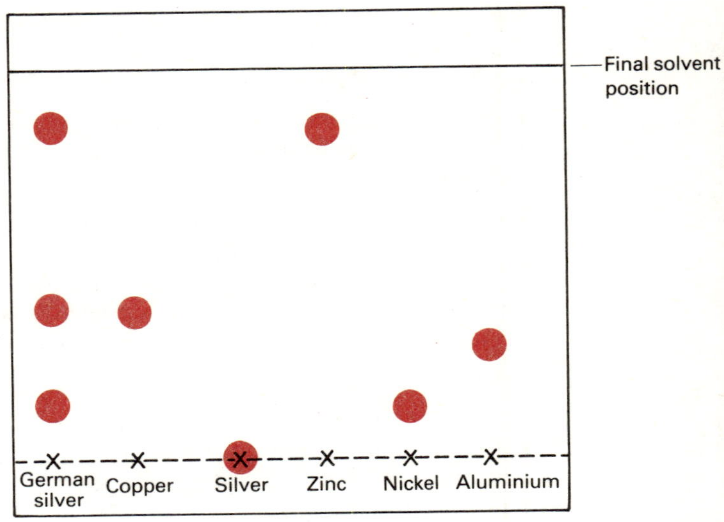

Fig. 14.2

Which of the substances is produced when

8 calcium is added to water

9 calcium carbonate is strongly heated

10 calcium oxide is 'slaked' by adding water

11 carbon dioxide is passed through limewater until the solution turns cloudy

12 carbon dioxide is passed through water until the solution goes clear again

13 molten calcium chloride is electrolyzed?

Questions 14-16

A lower melting point than pure metals
B poorer electrical conductivity than pure metals
C more resistant to corrosion than pure metals
D stronger than pure metals

Four properties of an alloy compared to its constituent pure metals are given in A–D above. Select the best reason for the use of the alloy rather than the pure metals in each of the following situations.
In these questions you may not choose any answer more than once.

14 for cutlery, e.g. stainless steel **15** for an aluminium alloy bicycle frame

16 for solder used to join metals together

Multiple completion

Type 1 (See instructions on page 1.)

17 Which of the following statements about the Kellner-Solvay process is (are) correct?
(*i*) The electrolyte is an aqueous solution of sodium chloride.
(*ii*) The products are sodium hydroxide, hydrogen and chlorine.
(*iii*) Mercury is used as the anode.

Type 2 (See instructions on page 1.)

18 Which of the following statements are true for **all** metals?
They
(*i*) conduct electricity. (*ii*) are denser than water. (*iii*) are silver in colour.
(*iv*) react with acids to form hydrogen.

Structured questions

19 Brass, copper, zinc, duralumin, steel, iron.
(*a*) What is meant by the term 'alloy'.
(*b*) Select those substances which are alloys from the above list.

The composition of an alloy called German silver can be obtained by experiment. A piece of German silver was added to concentrated nitric acid. A violent reaction took place and copious brown fumes were evolved. The solution remaining after the reaction was blue-green in colour. It was diluted with water. A spot of this solution was placed on a square of filter paper together with spots of other solutions of different metal nitrates in water. After drying the filter paper was placed in a large beaker with a small volume of solvent in the bottom. The solvent moved up the filter paper causing the spots on the filter paper to move. After drying, the spots were developed and the result is shown in Fig. 14.2.

(*c*) What are the brown fumes formed during the reaction of German silver with concentrated nitric acid?

(*d*) What precautions should be taken when carrying out this part of the experiment?

(*e*) What remains in the solution after the reaction?

(*f*) From the results in Fig. 14.2, what can be concluded about the metals present in German silver?

(*g*) What name is given to this technique? (*h*) Suggest one use of German silver.

15 Metals(II)

Simple multiple choice

1 A metal high in the reactivity series
A forms compounds which are unstable.
B is obtained by reduction of an ore with carbon.
C readily gains electrons.
D burns to form an oxide which has a pH below 7.
E reacts readily with cold water to form hydrogen.

2 A metal ore is best defined as a
A compound of a metal. B mixture of a metal and impurities.
C mineral containing compounds of a metal.
D natural form of a metal.

3 Which one of the following metals is most likely to be found native (i.e. chemically uncombined) in the ground?
A zinc B gold C calcium D iron E sodium

4 A metal is required which has a low density and is resistant to corrosion. Which one of the following metals is most suitable?
A sodium B aluminium C iron D lead E copper

5 Which one of the following is **not** deliberately added to the blast furnace in the extraction of iron?
A carbon dioxide B limestone C air D coke E iron ore

6 Limestone is added to the blast furnace in order to
A reduce the iron ore to iron. B act as a fuel.
C lower the temperature of the furnace.
D remove the impurities as slag.
E convert the coke into carbon dioxide.

7 Magnesium, zinc, lead and silver powders were added in excess to individual samples of blue copper(II) sulphate solution. How many solutions will still be blue after a few days? (Solutions of magnesium, zinc, lead and silver compounds are colourless.)
A 4 B 3 C 2 D 1 E 0

8 Chromium does not react with cold water but reacts with dilute hydrochloric acid to produce hydrogen. Lead is produced when chromium is dropped into lead(II) nitrate solution.
The order of reactivity of the three metals in order of decreasing reactivity is
A calcium, chromium, lead. C chromium, calcium, lead. E lead, chromium, calcium.
B calcium, lead, chromium. D chromium, lead, calcium.

9 The nitrates of three metals, P, Q and R, are heated in dry test tubes. The nitrate of P gives off oxygen only while the nitrates of Q and R give off oxygen and nitrogen dioxide. The nitrate of Q leaves a residue of a metal and that of R a metal oxide.
Which one of the following gives the order of increasing reactivity of the metals?
A P, Q, R B P, R, Q C Q, R, P D Q, P, R E R, Q, P

10 Various mixtures of four metals, W, X, Y and Z, and their oxides were heated with the following results:

X oxide + Z → X + Z oxide
W oxide + X → W + X oxide
Y oxide + W → Y + W oxide

The order of reactivity of the four metals with the most reactive first is
A X, Z, W, Y. C W, Y, X, Z. E Z, W, X, Y.
B Z, X, W, Y. D Y, X, W, Z.

11 The main reason for adding calcium chloride to sodium chloride in the electrolysis of molten sodium chloride is to

A absorb water which would otherwise react with the sodium formed.
B increase the conductivity of the electrolyte.
C act as a catalyst and speed up the reaction.
D lower the melting point of the electrolyte.

12 Three samples of steel were exposed to different atmospheric conditions for the same period of time:

1 in a laboratory where all water vapour is removed from the air
2 dipping into sea water on a harbour wall
3 exposed on the roof of a building in a town

Arrange the three samples, 1, 2 and 3, in order of the amount of corrosion which takes place putting the least corroded sample first. Code the answer according to the list A–E below.

A 1, 2, 3 B 2, 3, 1 C 1, 3, 2 D 3, 1, 2 E 2, 1, 3

Questions 13–15 refer to experiments carried out to measure the voltages produced by simple cells using the apparatus in Fig. 15.1.

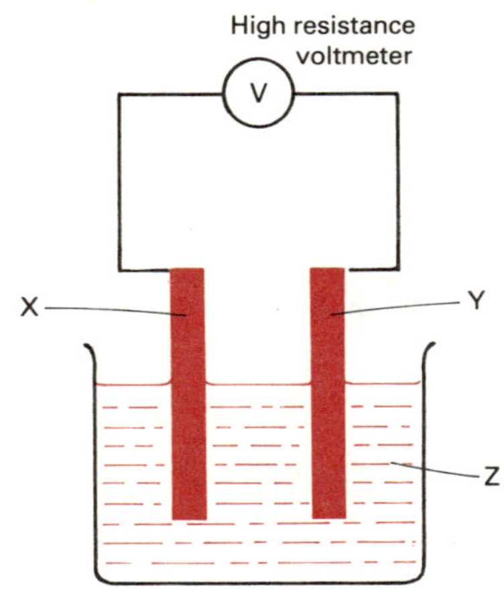

Fig. 15.1

Two metal rods X and Y are connected to a high resistance voltmeter and dipped into a liquid Z.

13 In which one of the following cases is a voltage recorded on the voltmeter?

A X and Y are made of the same metal and Z is pure water
B X and Y are different metals and Z is pure water
C X and Y are made of the same metal and Z is dilute sulphuric acid
D X and Y are different metals and Z is dilute sulphuric acid

14 Using salt solution for Z, the metals X and Y were changed. Which one of the following pairs of metals would give the largest reading on the voltmeter?

	X	Y
A	copper	copper
B	magnesium	magnesium
C	copper	silver
D	magnesium	copper
E	magnesium	silver

15 The series of experiments were repeated using four metals (labelled I, II, III and IV) in turn with an iron rod and a salt solution Z. The results obtained were

Metal used	*Voltage (V)*
I	−0.55
II	−0.35
III	+0.20
IV	−0.10

Using only this evidence, arrange the four metals in order of reactivity putting the most reactive metal first.

A I, II, IV, III
B III, IV, II, I
C II, III, IV, I
D IV, I, III, II
E III, II, I, IV

Classification

Questions 16–18

A calcium B copper C gold D aluminium E sodium

Which of the metals above is

16 extracted from the oxide by electrolysis?

17 most likely to be found uncombined in the earth?

18 displaced from solutions of its salts by all of the other metals?

Multiple completion

Type 1 (See instructions on page 1.)

19 Which of the following metals will displace zinc from zinc oxide?
(*i*) magnesium (*ii*) iron (*iii*) copper

20 Ores of metals near the top of the reactivity series are
(*i*) more difficult to split up than ores of metals lower in the series.
(*ii*) more likely to be chlorides and carbonates than oxides or sulphides.
(*iii*) usually converted to the metal by electrolysis.

Type 2 (See instructions on page 1.)

21 Which of the following tests would be useful in establishing the position of nickel in the reactivity series?
(*i*) the reaction of nickel with water or steam
(*ii*) the addition of nickel to copper(II) sulphate solution
(*iii*) heating nickel(II) oxide in hydrogen
(*iv*) the addition of sodium hydroxide solution to a solution of nickel(II) sulphate.

22 Which of the following compounds, on heating in the laboratory, do **not** decompose to form an oxide?
(*i*) copper(II) carbonate
(*ii*) copper(II) sulphate
(*iii*) lead(II) nitrate
(*iv*) sodium carbonate

23 Which of the following mixtures when heated will react to form new substances?
(*i*) copper and magnesium oxide
(*ii*) iron and zinc oxide
(*iii*) magnesium and copper(II) oxide
(*iv*) zinc and calcium oxide

Assertion-reason

(See instructions on pages 1–2.)

24 Zinc is used to protect iron from rusting	Iron is a more reactive metal than zinc.
25 No reaction occurs when a piece of lead is dipped into zinc sulphate solution.	Lead is below zinc in the reactivity series.

Structured questions

26 Aluminium is the most abundant metal in the earth's crust. It makes up 7.5% compared to 4.7% for iron. Aluminium is a constituent of all clays.
(*a*) Why is aluminium not extracted from clay?
(*b*) Why is aluminium much more expensive than iron although it is more abundant than iron?

Aluminium is manufactured by the electrolysis of molten aluminium oxide obtained from bauxite $Al_2O_3.2H_2O$. The bauxite is added to sodium hydroxide solution and the impurities remain undissolved. The impurities are removed and aluminium oxide precipitated. The aluminium is removed and heated strongly to produce aluminium oxide.

The melting point of aluminium oxide is 2045°C and the electrolysis of molten aluminium oxide is uneconomic. The aluminium is dissolved in molten cryolite (sodium aluminium fluoride). The temperature of the molten mixture is below 1000°C. The products of the electrolysis are aluminium (at the cathode) and oxygen (at the anode).

(*c*) Aluminium oxide dissolves in both alkalis and acids. What name is given to oxides which dissolve in both acids and alkalis?
(*d*) Name a procedure for removing insoluble impurities from the solution.
(*e*) Why is there a sloping floor to the electrolysis cell?
(*f*) Why is the electrolysis of molten aluminium oxide not economic?
(*g*) Write ionic equations to show the changes taking place at each electrode.
(*h*) Why do the carbon anodes have to be regularly replaced?
(*i*) Give one reason why aluminium extraction factories are built in mountainous areas.
(*j*) Aluminium and its alloys have a wide range of uses. For each of the following uses give an advantage of using aluminium or an aluminium alloy. Give a different advantage in each case.
(*i*) overhead power cables (*iii*) kitchen foil (*v*) bicycle frames
(*ii*) panels on an aircraft wing (*iv*) saucepans
(*k*) Modern warships have considerable amounts of aluminium in the superstructure.
(*i*) Why is aluminium used rather than steel?
(*ii*) What is the disadvantage of using aluminium if the ship is struck by a missile?

16 Oxidation and reduction

There are also other examples of questions on oxidation and reduction throughout the book.

Simple multiple choice

1 Which one of the following statements contains three correct descriptions of oxidation?
A addition of oxygen, addition of hydrogen, addition of electrons
B addition of oxygen, loss of hydrogen, addition of electrons
C addition of oxygen, loss of hydrogen, loss of electrons
D addition of oxygen, addition of hydrogen, loss of electrons
E loss of oxygen, loss of hydrogen, loss of electrons

2 An oxidizing agent
A always contains oxygen. B is readily oxidized.
C is oxidized only in the presence of a reducing agent. D is reduced during reactions.

3 Which one of the following reactions does **not** represent an oxidation?
A $2Fe + 3Cl_2 \rightarrow 2FeCl_3$
B $2Cl^- \rightarrow Cl_2 + 2e^-$
C $Cl_2 + 2I^- \rightarrow 2Cl^- + I_2$
D $2FeCl_2 + Cl_2 \rightarrow 2FeCl_3$
E $Ag^+ + Cl^- \rightarrow AgCl$

4 When iron(II) sulphate is left exposed to the air it turns brown. This is because
A iron(II) ions are oxidized.
B rusting takes place.
C reduction takes place.
D iron(II) ions gain electrons.

5 Which one of the following is **not** an example of oxidation?
A $Mg \rightarrow Mg^{2+}$
B $S^{2-} \rightarrow S$
C $I_2 \rightarrow 2I^-$
D $Mn^{2+} \rightarrow MnO_4^-$

Multiple completion

Type 1 (See instructions on page 1.)

6 Common reducing agents include (*i*) hydrogen (*ii*) carbon monoxide (*iii*) chlorine.

7 Which of the following reagents could be used to convert an aqueous solution containing Fe^{2+} ions to a solution of Fe^{3+} ions?
(*i*) dilute nitric acid. (*ii*) sodium hydroxide solution. (*iii*) ammonia solution.

8 Which of the following equations represent redox reactions?
(*i*) $FeSO_4(aq) + 2NaOH(aq) \rightarrow Fe(OH)_2(s) + Na_2SO_4(aq)$
(*ii*) $KOH(aq) + HCl(aq) \rightarrow KCl(aq) + H_2O(l)$
(*iii*) $CuSO_4(aq) + Zn(s) \rightarrow ZnSO_4(aq) + Cu(s)$

9 In which of the following reactions is sulphuric acid acting as an oxidizing agent?
(*i*) $Cu(s) + 2H_2SO_4(l) \rightarrow CuSO_4(aq) + 2H_2O(l) + SO_2(g)$
(*ii*) $Cu(OH)_2(s) + H_2SO_4(aq) \rightarrow CuSO_4(aq) + 2H_2O(l)$
(*iii*) $CuO(s) + H_2SO_4(aq) \rightarrow CuSO_4(aq) + H_2O(l)$

Type 2 (See instructions on page 1.)

10 In which of the following is the substance in bold type acting as an oxidizing agent?
(*i*) **Chlorine** is bubbled through potassium bromide solution.
(*ii*) **Magnesium** is added to dilute hydrochloric acid.
(*iii*) **Manganese(IV) oxide** is added to concentrated hydrochloric acid.
(*iv*) **Manganese(IV) oxide** is added to hydrogen peroxide

Structured questions

11 For each of the following conversions
(*i*) name a suitable reagent for carrying out the reaction.
(*ii*) state whether the reaction involves the oxidation of the substance in bold, the reduction of the substance in bold or whether the substance in bold is neither oxidized nor reduced.
(*iii*) write a balanced equation for the reaction.
(*a*) **sodium carbonate** to carbon dioxide (*b*) **iron(II) chloride** to iron(III) chloride
(*c*) **ammonia** to nitrogen (*d*) **ammonium** chloride to ammonia
(*e*) **potassium iodide** to iodine

17 The effect of electricity on chemicals

Simple multiple choice

1 An electrolyte is best defined as a compound which
A contains a metal. B only conducts electricity when liquid or in aqueous solution.

C conducts electricity under all conditions.
D conducts electricity only when dissolved in water.

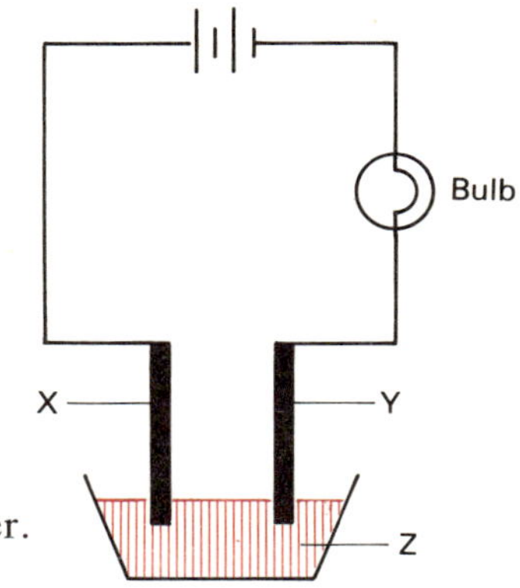

Fig. 17.1

2 Which one of the following does **not** require the use of electricity?
A the galvanizing of iron
B the extraction of aluminium from bauxite
C the electroplating of steel D the anodizing of aluminium

3 A substance which conducts electricity when molten but not when solid is
A an ionic compound. C a metallic element.
B a covalent compound. D a non-metallic element.

4 Lead can be obtained from lead(II) bromide crystals by
A the action of heat. B heating a mixture of lead(II) bromide and copper.
C splitting open the crystals. D electrolysis of molten lead(II) bromide.

Questions 5–7

The apparatus in Fig 17.1 was used to test substances to find out if they conducted electricity. The following substances were used (labelled Z in the diagram): molten wax, mercury, molten sodium chloride, lead, distilled water and a dilute aqueous solution of hydrochloric acid.

5 In how many of these tests would the bulb light?
A 1 B 2 C 3 D 4 E 5

6 The experiment was carried out using copper(II) sulphate solution as Z. The bulb lit up. Which one of the following observations would also be made?
A Bubbles of gas are produced at the cathode.
B A brown deposit of copper forms on the anode.
C Bubbles of gas are produced at the anode. D The temperature of the solution rises.

7 Which one of the following would cause the bulb to glow brighter but not speed up the rate of electrolysis of copper(II) sulphate solution?
A move X and Y closer together C increase the current flowing
B connect X and Y with a copper wire D push X and Y deeper into Z

8 When electricity was passed through a concentrated solution of hydrochloric acid, it was found that
A chlorine was formed at the anode. C hydrogen was formed at the anode.
B hydrogen chloride gas was produced. D no gases were formed.
E chlorine was formed at the cathode.

9 Electrolysis of a colourless aqueous solution using carbon electrodes produced a brown colour around one electrode and a metal was deposited on the other. The substance present in the aqueous solution could be
A copper(II) bromide. C sodium bromide. E potassium iodide.
B zinc bromide. D silver iodide.

10 What are the main products of the electrolysis of an aqueous solution of potassium nitrate using carbon electrodes?
A potassium and nitrogen C potassium and hydrogen
B potassium, nitrogen and oxygen D hydrogen and nitrogen
E hydrogen and oxygen

11 In the electrolysis of molten lead(II) bromide, which of the equations below represents the reaction occurring at the anode (positive electrode)?
A $Pb \rightarrow Pb^{2+} + 2e^-$ C $Br^- + e^- \rightarrow Br$ E $Br + e^- \rightarrow Br$
B $Pb^{2+} + 2e^- \rightarrow Pb$ D $Br^- \rightarrow Br + e^-$

12 When the electrolysis of molten calcium bromide is carried out, the reaction taking place at the cathode is
A $Ca^{2+} + 2e^- \rightarrow Ca$. C $2Br^- - 2e^- \rightarrow Br_2$.
B $Ca^{2+} - 2e^- \rightarrow Ca$. D $2Br^- + 2e^- \rightarrow Br_2$.

13 Electrolysis of molten lithium hydride (LiH) produces hydrogen at the anode because
A the anode is negatively charged. C lithium hydride contains H^+ ions.
B lithium hydride contains H^- ions. D hydrogen is more reactive than lithium.

14 An object, made of copper, is silver plated by making it the cathode in an electrolysis process. The anode and the electrolyte are
A copper and silver nitrate solution.
B silver and silver nitrate solution.
C silver and copper(II) sulphate solution.
D carbon and silver nitrate solution. E silver and dilute nitric acid.

Questions 15 and 16

During an electrolysis of a metal sulphate solution, 0.5 g of metal were deposited when a current of 0.5 A flowed for 15 minutes.

15 How many faradays of electricity were used?

A $0.5 \times 15 \times 96\,500$

B $0.5 \times 15 \times 60 \times 96\,500$

C $\frac{0.5 \times 15}{96\,500}$

D $\frac{0.5 \times 15 \times 60}{96\,500}$

16 How many grams of metal would be deposited from the same solution if a current of 1 A flowed for 30 minutes?

A 0.5 g B 1.0 g C 1.5 g D 2.0 g E None of these.

17 The same quantity of electricity was passed through solutions of copper(II) sulphate and the nitrate of a metal M. 1.6 g of copper and 5.4 g of M were deposited. Which one of the following represents the ion of M? (A_r(Cu) = 64, A_r(M) = 108)

A M^+ B M^{2+} C M^{3+} D M^- E M^{2-}

18 Which one of the following arrangements below would be used in order to electroplate an object with nickel?

	Anode	*Cathode*	*Electrolyte*
A	object	nickel	nickel(II) sulphate solution
B	nickel	object	nickel(II) sulphate solution
C	nickel	object	dilute sulphuric acid
D	object	nickel	dilute sulphuric acid
E	carbon	object	nickel(II) sulphate solution

Questions 19–23

The apparatus in Fig. 17.2 was used to investigate the electrolysis of molten lead(II) bromide. The products of the electrolysis are lead and bromine.

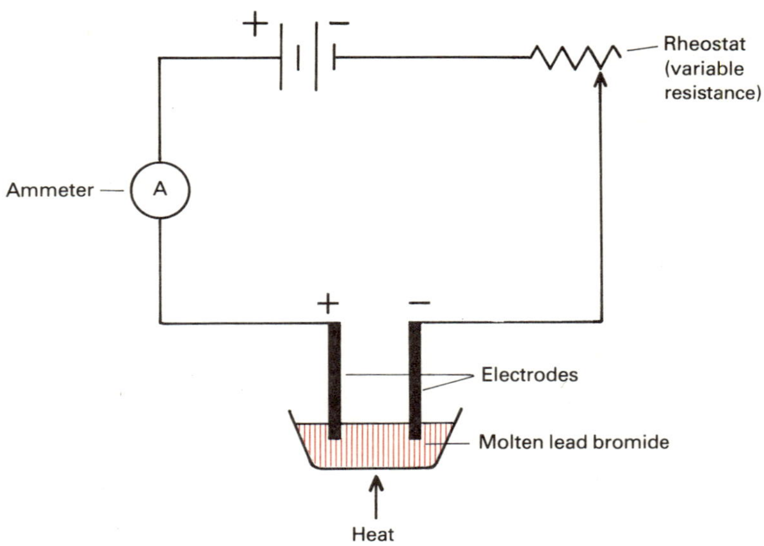

Fig. 17.2

19 Electricity is carried through the molten lead(II) bromide because of the presence of

A a metal. B free electrons. C free ions. D free atoms. E free molecules.

20 The lead(II) bromide in this experiment is called the

A electrolyte. B electrode. C cation. D conductor.

21 The purpose of the ammeter is to

A show that a current is flowing. B adjust the current flowing.

C ensure that alternating current is converted into direct current.

D measure the current flowing. E measure the amount of electricity passed.

22 What quantity of electricity is passed when a current of 5 amperes flows for 193 seconds? (1 faraday = 96 500 coulombs)

A 5×193 faradays

B $5 \times 193 \times 96\,500$ faradays

C $5 \times 193 \times 96\,500$ coulombs

D $\frac{5 \times 193}{96\,500}$ coulombs

E $\frac{5 \times 193}{96\,500}$ faradays

23 The amount of lead produced by the passage of 0.2 Faraday is
A 0.01 mole of lead atoms.
B 0.02 mole of lead atoms.
C 0.05 mole of lead atoms.
D 0.1 mole of lead atoms.
E 0.2 mole of lead atoms.

Multiple completion

Type 1 (See instructions on page 1.)

24 During the electrolysis of an aqueous solution of copper(II) sulphate using copper electrodes
(*i*) the copper anode decreases in mass.
(*ii*) copper is deposited on the cathode.
(*iii*) the colour of the electrolyte changes.

25 During the electrolysis of dilute sulphuric acid using platinum electrodes
(*i*) hydrogen is liberated at the cathode.
(*ii*) oxygen is liberated at the anode.
(*iii*) equal volumes of hydrogen and oxygen are liberated at the two electrodes.

Assertion-reason

(See instructions on pages 1–2.)

	Assertion	Reason
26	Molten calcium bromide undergoes electrolysis.	Calcium is a metal and metals conduct electricity.
27	Chloride, hydroxide and iodide ions all migrate towards the positive electrode during electrolysis.	Chloride, hydroxide and iodide ions are all negatively charged.

Structured questions

28 (*a*) (*i*) What name is given to the process used to coat one metal with a thin coating of another by means of electrolysis?
(*ii*) Describe how a copper disc can be coated with a thin layer of nickel by means of electrolysis.
(*iii*) If the electrolysis is carried out too quickly the nickel coating is inclined to flake off the copper disc. Give two ways of greatly slowing down the rate of electrolysis.

(*b*) When pieces of nickel foil are dropped into aqueous solutions of copper(II) sulphate and iron(II) sulphate, the piece of nickel foil dropped into iron(II) sulphate solution is unchanged but the piece of nickel foil dropped into copper(II) sulphate solution is coated with a brown coating.
(*i*) Name the brown coating formed on the nickel foil in a solution of copper(II) sulphate solution.
(*ii*) What information do these experiments give about the position of nickel compared to copper and iron in the reactivity series?

(*c*) Bicycle handlebars are made of steel. The steel tubing is coated with a thick coating of nickel and a very thin coating of another metal.
(*i*) What is the reason for nickel plating the steel?
(*ii*) Name the metal used to plate the nickel and explain why this metal is better than nickel for handlebars.

18 Chemical families and the Periodic Table

Simple multiple choice

1 At room temperature (20°C) and pressure, iodine is a
A colourless gas.
B red liquid.
C purple vapour.
D purple liquid.
E dark grey solid.

2 The halogens are placed in the same group of the Periodic Table because they
A are all gases at room temperature.
B all have seven electrons in the outer shell.
C all contain the same number of protons.
D all form compounds with hydrogen.
E are all non metallic elements.

3 Caesium belongs to the same family of elements as sodium and potassium (i.e. alkali metal family). Which one of the following is true?
Caesium
A is stored under water.
B burns in oxygen to form an acidic oxide.
C burns in chlorine to form a solid chloride
D is obtained by the electrolysis of an aqueous solution of caesium chloride.

4 Which one of the noble gases is used for filling balloons?
A helium. B neon C argon D krypton E xenon

5 The elements in the Periodic Table are arranged in order of their
A dates of discovery. C atomic sizes. E melting and boiling points.
B atomic numbers. D reactivities with water.

6 The horizontal rows in the Periodic Table are called
A triads. B groups. C periods. D families.

7 The formula of rubidium fluoride is
A RbF_7. B Rb_7F. C RbF. D RbF_2. E Rb_2F_3.

8 How many periods are required in the Periodic Table to accommodate the first eighteen elements?
A 1 B 2 C 3 D 7 E 8

9 Francium, Fr, is a radioactive metal in group I of the Periodic Table. Which one of the following statements about francium is correct?
Francium
A forms an insoluble chloride.
B forms a carbonate with the formula $FrCO_3$.
C reacts violently with cold water. D burns to form an acidic oxide.

10 Which one of the following best explains the position of carbon in the middle of the Periodic Table?
Carbon
A forms a very large number of compounds with hydrogen.
B forms two gaseous oxides. C can exist in two allotropic forms.
D does not form positive or negative ions. E is a non metallic element.

11 The following equation summarizes the reaction between chlorine and a solution of potassium iodide.

$$2KI(aq) + Cl_2(g) \rightarrow 2KCl(aq) + I_2(aq)$$

Which one of the following would react with an aqueous solution of potassium bromide to produce bromine?
A chlorine B potassium chloride C iodine D potassium iodide

12 Which one of the following would be most likely to form ions which, when dissolved in water, would give coloured solutions?
A group I elements C noble gases E group VII elements
B transition elements D group II elements

13 Which pair of elements would react together most vigorously?
A lithium and fluorine C potassium and fluorine E sodium and chlorine
B magnesium and chlorine D lithium and iodine

Classification

Questions 14–20

Table 18.1 shows part of the Periodic Table with five of the elements replaced by the letters A–E. These are not the usual symbols for these elements and you are not expected to identify them.

Period \ Group	I	II	III	IV	V	VI	VII	0
1								
2							E	
3		B	C	D				
4	A							
5								
6								

Table 18.1

Which one of these elements

14 would form ions with a 3+ charge?

15 is most likely to be a gas?

16 is likely to be a non metal with a 'diamond-like' structure?

17 contains the greatest number of protons in each atom?

18 would contain the greatest number of atoms in a 100 g sample of each element?

19 would form an oxide in which two moles of atoms of the element combine with one mole of atoms of oxygen?

20 is in the same period of the Periodic Table as carbon?

Questions 21–24
From the following five groups in the Periodic Table choose the group in which each of the elements W, X, Y and Z belongs.

A group I B group II C group III D group IV E group VI

21 W is an element which conducts electricity and melts below 100°C.

22 X is a yellow solid which combines with hydrogen to form a hydride with a formula H_2X.

23 Y is a metal which burns in oxygen to form an oxide YO.

24 Z is a metalloid. It is extracted from an oxide by heating with magnesium. It forms a chloride with a formula ZCl_4.

Multiple completion

Type 1 (See instructions on page 1.)

25 When moving across any period of the Periodic Table from group I to group VII
(*i*) the properties of the elements change from metallic to non metallic.
(*ii*) the oxides change from basic to acidic.
(*iii*) there is a steady increase in atomic number.

26 Which of the following properties are common to the three elements calcium, potassium and sodium?
(*i*) they are all metals. (*ii*) they form ions by losing one electron.
(*iii*) they are in the same group of the Periodic Table.

27 Noble gases
(*i*) undergo no chemical reactions.
(*ii*) are obtained by thermal decomposition of minerals.
(*iii*) increase in boiling point down the group.

Type 2 (See instructions on page 1.)

28 Elements in the same group
(*i*) have similar chemical properties. (*ii*) contain the same number of electrons.
(*iii*) have the same number of electrons in the outer shell.
(*iv*) contain the same number of protons.

Assertion-reason

(See instructions on pages 1–2.)

29 Carbon and silicon both form acidic oxides. Carbon and silicon are in the same period of the Periodic Table.

30 Bromine liberates iodine from a solution of potassium iodide. Bromine is more reactive than iodine.

Structured questions

31 The following passage concerns the chemistry of the element boron. Read the passage carefully and use it to answer questions (*a*)–(*g*).

Boron (symbol B) is a non metallic element which was first prepared in 1808 by J. L. Gay Lussac. He prepared it by heating boron(III) oxide with metallic potassium. It can also be prepared by the reduction of boron(III) oxide with magnesium. This reduction is very similar to the reduction of silicon(IV) oxide with magnesium to produce silicon.

The element boron occurs in the earth's crust as boric(III) acid and as borates. It makes up about 0.001 per cent of the earth's crust.

Pure boron can exist in both crystalline and powder forms. The crystalline form is produced by heating a tantalum filament in an atmosphere of hydrogen and boron(III) fluoride vapour. In the crystalline form boron exists in a three-dimensional giant structure with strong bonds between atoms.

Powdered boron is more reactive than crystalline boron. Boron acts as a reducing agent at high temperatures. It reduces many metal oxides producing boron(III) oxide. Often the metal obtained reacts with boron to form borides.

Boron(III) chloride is prepared by the action of chlorine gas on heated boron. It has a boiling point of 12°C and is a non conductor of electricity.

Compounds of boron with hydrogen are called boranes. They are produced when dilute hydrochloric acid is added to the residue remaining from the reaction between boron(III) oxide and magnesium. The best known borane is diborane B_2H_6.

Boron has an atomic number of 5.

(*a*) (*i*) Write the symbol equation for the reaction of boron(III) oxide and magnesium.
(*ii*) How could a sample of boron be obtained from the residue?

(*b*) (*i*) What is the function of the tantalum filament in the production of crystalline boron?
(*ii*) How could the filament be heated?
(*iii*) What is the action of hydrogen in the reaction?
(*iv*) Write a symbol equation for the reaction occurring on the filament.

(*c*) (*i*) Write a symbol equation for the preparation of boron(III) chloride.
(*ii*) What type of bonding exists in boron(III) chloride?

(*d*) Boranes fit a general formula B_nH_{n+4}. Write the molecular formula of the borane containing five boron atoms. Which other element forms a wide range of hydrides?

(*e*) Write the symbol equation for the reaction of magnesium boride (Mg_3B_2) with dilute hydrochloric acid to produce diborane.

(*f*) What is the electron arrangement in a boron atom?

(*g*) In which group and period of the Periodic Table is boron placed?

19 Acids, bases and salts

Simple multiple choice

1 An alkali is a base which
A is decomposed by heating.
B forms acid salts.
C is neutral.
D dissolves in water.

2 A colourless gas was dissolved in water and produced an alkaline solution. The gas could be
A ammonia.
B carbon dioxide.
C sulphur dioxide.
D hydrogen chloride.
E methane.

3 A soil with a pH less than 5 can be made less acid by adding
A ammonium sulphate.
B calcium phosphate.
C calcium hydroxide.
D ammonium nitrate.

4 An insoluble salt is best prepared by
A reduction.
B neutralization.
C precipitation.
D decomposition.
E direct combination.

5 If some **dry** tartaric acid crystals are placed on **dry** blue litmus paper the paper will
A turn red.
B remain unchanged.
C turn purple.
D bleach (i.e. go white).
E turn yellow.

6 The reaction between zinc oxide and dilute hydrochloric acid produces zinc chloride and also
A hydrogen chloride.
B hydrogen.
C water.
D carbon dioxide.
E zinc hydroxide.

7 A solution with a pH of 10 can be neutralized by adding
A distilled water.
B universal indicator.
C ethanol.
D ethanoic acid.
E sodium hydrogencarbonate.

8 Which one of the following equations represents a neutralization?
A $2H_2(g) + O_2(g) \rightarrow 2H_2O(l)$
B $NaOH(aq) + HNO_3(aq) \rightarrow NaNO_3(aq) + H_2O(l)$
C $2NaOH(aq) + FeSO_4(aq) \rightarrow Fe(OH)_2(s) + Na_2SO_4(aq)$
D $PbO(s) + H_2(g) \rightarrow Pb(s) + H_2O(l)$
E $Zn(s) + H_2SO_4(aq) \rightarrow ZnSO_4(aq) + H_2(g)$

9 Magnesium(II) sulphate can be made by reacting hot, dilute sulphuric acid and **excess** magnesium oxide?

Which one of the following sets of directions would give the purest product?
A evaporate until saturated, cool, filter
B cool, evaporate until saturated, decant
C evaporate until saturated, cool, decant
D filter, evaporate until saturated, cool

10 A large excess of powdered calcium carbonate was added to an aqueous solution of pH 2. At the end of the reaction the pH would be
A less than 2.
B 2.
C between 2 and 6.
D 7.
E above 7.

11 Three liquids were available:
I a solution of dry hydrogen chloride in dry methylbenzene;
II pure ethanol; III pure water.
Which one of the following would have a pH less than 7?
A I B II C I and II D I and III E II and III

12 Lead(II) sulphate is best prepared by adding a dilute solution of sulphuric acid to
A lead. B powdered lead(II) oxide. C powdered lead(II) carbonate.
D an aqueous solution of lead(II) nitrate. E powdered lead(II) nitrate.

13 After precipitating barium sulphate the precipitate is removed by filtration and warm water is poured onto the filter paper. The warm water is added in order to
A dissolve the barium sulphate.
B neutralize any unused acid present.
C remove dirt from the precipitate.
D remove traces of the reactants.
E speed up the drying process.

14 Which one of the following is **not** a property of an acid solution?
An acid solution
A has a sharp taste. B has a pH more than 7.
C produces carbon dioxide with carbonates.
D neutralizes alkalis. E reacts with metal oxides.

15 When a proton is lost from an acid the resulting particle is called the conjugate base. Which one of the following is the conjugate base of the hydrogensulphate ion HSO_4^-?
A SO_4^{2-} B H_2SO_4 C SO_3 D HSO_4 E SO_3^{2-}

16 In which one of the following will ethanoic acid have the highest electrical conductivity?
A in the vapour state
B in the liquid state
C in an aqueous solution
D dissolved in an organic solvent
E in the solid state

Classification

Questions 17–22
Five solutions labelled A–E had the following pH values.
A pH 1 B pH 6 C pH 7 D pH 9 E pH 12
Select A–E from the above list to answer the following questions.
Which of the solutions

17 is exactly neutral? **18** is strongly acidic? **19** is weakly alkaline?

20 could be a solution of carbon dioxide in water?

21 would be required in the smallest volume to neutralize a solution of potassium hydroxide?

22 should be added to some dilute sulphuric acid spilt on the laboratory floor?

Questions 23 and 24
The following equations all involve water.
A $H_2O(g) + Zn(s) \rightleftharpoons ZnO(s) + H_2(g)$
B $H_2O(l) + NH_3(g) \rightleftharpoons NH_4^+(aq) + OH^-(aq)$
C $H_2O(l) + SO_3(g) \rightarrow H_2SO_4(l)$
D $H_2O(l) + HCl(g) \rightarrow H_3O^+(aq) + Cl^-(aq)$
E $2H_2O(l) + 2Na(s) \rightarrow 2Na^+(aq) + 2OH^-(aq) + H_2(g)$
In which of the above equations is water acting as

23 an acid? **24** a base?

Multiple completion

Type 1 (See instructions on page 1.)

25 Which of the following statements is (are) true for both 0.1 M sodium hydroxide and 0.1 M ammonia solution?
Both solutions
(*i*) have a pH greater than 7. (*ii*) turn universal indicator the same colour.
(*iii*) contain equal numbers of hydroxide ions.

26 A solution containing a large concentration of $OH^-(aq)$
(*i*) has a pH greater than 7. (*ii*) precipitates metal hydroxides from salt solutions.
(*iii*) conducts electricity.

27 A white precipitate is formed when two solutions are mixed. The two solutions could be
(*i*) potassium nitrate and sodium carbonate.
(*ii*) silver nitrate and sodium chloride. (*iii*) lead(II) nitrate and sodium carbonate.

Type 2 (See instructions on page 1.)

28 Which of the following ions are present in an aqueous solution of ammonium chloride?
(*i*) $NH_4^+(aq)$ (*ii*) $OH^-(aq)$ (*iii*) $Cl^-(aq)$ (*iv*) $H^+(aq)$

29 Which of the following would react with dilute sulphuric acid producing magnesium sulphate?
(*i*) magnesium
(*ii*) magnesium oxide
(*iii*) magnesium carbonate
(*iv*) magnesium chloride

30 Which of the following would react with an acid to produce a soluble salt?
(*i*) a base (*ii*) an alkali (*iii*) a reactive metal (*iv*) a carbohydrate

Structured questions

31 (*a*) Complete the following table which shows various methods of preparing salts.

Names of reacting substances	*Name of salt prepared*	*Names of other substances produced*
Sodium hydroxide + Nitric acid	**A**	**B**
Zinc + **C**	Zinc sulphate	**D**
Sodium sulphate + **E**	Barium sulphate	**F**
G + **H**	Iron(III) chloride	—
I + **J**	Lead(II) nitrate	Carbon dioxide and water

Table 19.1

Copper(II) sulphate ($CuSO_4$) can be prepared by the reaction between copper(II) oxide (CuO) and dilute sulphuric acid (H_2SO_4). Some dilute sulphuric acid is heated to about 60°C. Copper(II) oxide is added in small portions to the acid. After each addition the solution is stirred. When some copper(II) oxide remains unchanged in the acid, the copper(II) oxide is removed.

The solution is then carefully evaporated but not to dryness. A glass rod is dipped into the solution from time to time. When crystals form on the end of the glass rod on cooling, the solution is ready to leave to cool. Crystals form on cooling.

(*b*) Write a balanced equation for the reaction forming copper(II) sulphate.
(*c*) Name the piece of apparatus most suitable for
(*i*) measuring the temperature of the solution.
(*ii*) adding the copper(II) oxide.
(*iii*) containing the solution during evaporation.
(*d*) Why does copper(II) oxide remain unchanged in the solution?
(*e*) Name a process for removing this copper(II) oxide from the solution.
(*f*) What colour change occurs when black copper(II) oxide is added to the sulphuric acid?
(*g*) Why should the copper(II) sulphate solution not be heated to dryness?

20 Rates of reaction

Simple multiple choice

1 A catalyst is a substance which
A will start a reaction off. B speeds up an already proceeding reaction.
C causes a reaction to become reversible.
D stops a reaction.
E causes a reaction to evolve energy.

2 Catalysts are widely used in industry because they
A remove impurities from reaction mixtures.
B raise the temperature of reaction mixtures.
C speed up the chemical reactions taking place.
D increase the amounts of the products which can be formed.
E remove the products from the reaction mixtures.

3 Zinc reacts with dilute sulphuric acid to produce hydrogen. The reaction is catalyzed by copper. Which one of the following, if added to 50 cm^3 of 2 M sulphuric acid, would produce 50 cm^3 of hydrogen in the shortest time?
A 5 g of lumps of zinc
B 5 g of lumps of zinc plus 0.5 g of copper powder
C 5 g of powdered zinc
D 5 g of powdered zinc plus 0.5 g of copper powder

4 In many chemical reactions a 10° temperature rise
A doubles the rate of reaction.
B doubles the time required for the reaction to be completed.
C halves the rate of reaction.
D speeds the rate of reaction slightly.
E has no effect on the rate of reaction.

5 Silver chloride, when precipitated by mixing aqueous solutions of silver nitrate and hydrochloric acid, is initially a white precipitate but turns purple on standing. The reason for this change is that the precipitate
A reacts with oxygen in the air.
B is decomposed by light.
C reacts with excess dilute hydrochloric acid.
D reacts with excess silver nitrate solution.

6 Manganese(IV) oxide is a catalyst for the decomposition of hydrogen peroxide. Which one of the following statements explains why the rate at which oxygen is produced decreases during the reaction?
A Manganese(IV) oxide is used up.
B The surface area of manganese(IV) oxide decreases.
C The solution of hydrogen peroxide becomes more dilute.
D The reaction is endothermic.
E Oxygen is used up as the reaction is reversible.

Questions 7–10

The reaction between calcium carbonate and dilute hydrochloric acid is represented by the equation:

$$CaCO_3(s) + 2HCl(aq) \rightarrow CaCl_2(aq) + CO_2(g) + H_2O(l)$$

1 g samples of calcium carbonate were added to 100 cm^3 of M hydrochloric acid (an excess) and the carbon dioxide produced collected at room temperature and pressure. The final volume of carbon dioxide collected was 240 cm^3.

7 At the end of each experiment
A a colourless solution remained.
B a coloured solution remained.
C a coloured precipitate and a colourless solution remained.
D a white precipitate and a coloured solution remained.
E some unreacted calcium carbonate remained.

8 The rate of reaction is increased by
A increasing the size of the reaction vessel.
B using smaller lumps of calcium carbonate.
C decreasing the temperature.
D using a smaller mass of calcium carbonate.
E using a smaller volume of M hydrochloric acid.

9 The volume of carbon dioxide produced would be exactly 480 cm^3 if
A the experiment was carried out at 30°C instead of 20°C.
B 1 g of finely powdered calcium carbonate was used.
C 2 g of calcium carbonate were used.
D 50 cm^3 of 2 M hydrochloric acid were used.

10 If the experiment is carried out using 100 cm^3 of M sulphuric acid, very little carbon dioxide is collected. This is because
A the carbon dioxide produced is very soluble in dilute sulphuric acid.
B sulphuric acid is less concentrated than hydrochloric acid.
C sulphuric acid is a weaker acid than hydrochloric acid.
D calcium sulphate is much less soluble than calcium chloride.

11 The mass of a catalyst was determined at intervals during a reaction. Which one of the four graphs in Fig 20.1 (page 42) would be obtained?

12 A piece of chalk and some powdered chalk, both weighing 5 g, were placed in separate beakers containing excess dilute hydrochloric acid. After two minutes no powdered chalk remained but the piece of chalk was still reacting. This is because powdered chalk
A is more soluble in acid.
B contains fewer particles.
C has a larger surface area.
D contains a catalyst.

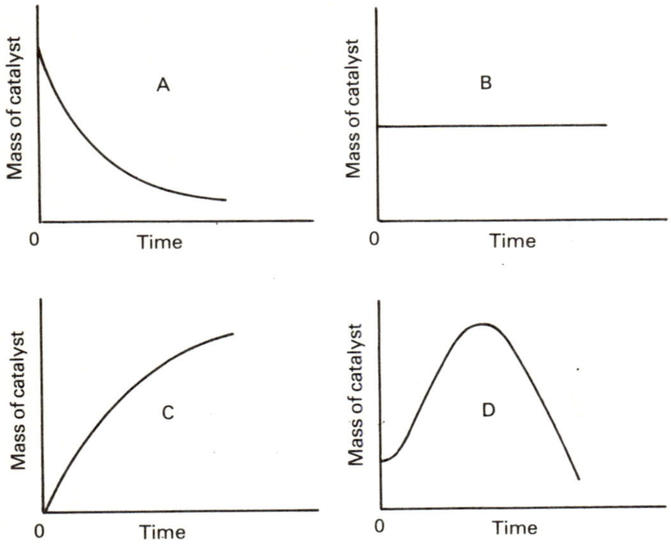

Fig. 20.1

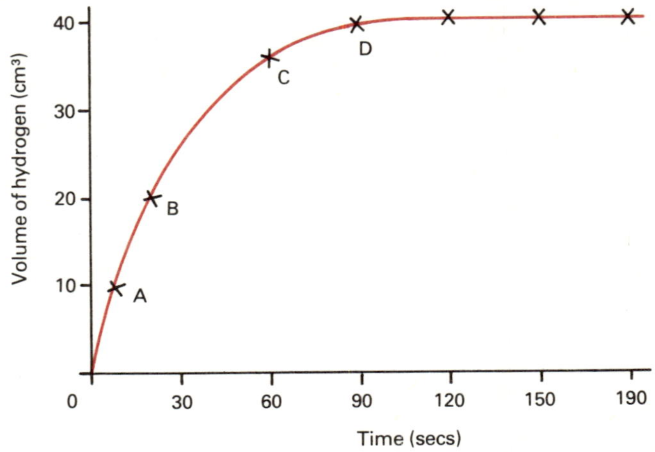

Fig. 20.2

13 Which one of the following reactions is affected by light?
A $H_2(g) + Cl_2(g) \rightarrow 2HCl(g)$
B $N_2(g) + 3H_2(g) \rightleftharpoons 2NH_3(g)$
C $NH_3(g) + HCl(g) \rightleftharpoons NH_4Cl(s)$
D $S(s) + O_2(g) \rightarrow SO_2(g)$

Questions 14–16

The reaction between sodium thiosulphate solution and dilute hydrochloric acid is represented by the following equation.

$$Na_2S_2O_3(aq) + 2HCl(aq) \rightarrow 2NaCl(aq) + SO_2(g) + S(s) + H_2O(l)$$

A series of experiments were carried out to show the effect on the rate of reaction of diluting the sodium thiosulphate solution.

14 The course of the reaction is followed by timing from the mixing of the solutions until
A gas bubbles are no longer escaping.
B the solution changes colour.
C the solution goes very cloudy.
D 100 cm^3 of gas are collected.

15 Which one of the following should be varied during this series of experiments?
A the temperature of the experiments
B the concentration of dilute hydrochloric acid
C the volume of dilute hydrochloric acid
D the concentration of sodium thiosulphate solution
E the volume of sodium thiosulphate solution

16 Which one of the following graphs would be most useful in establishing that the rate of reaction is proportional to the concentration (c) of sodium thiosulphate solution? (T represents time.)
A c against T
B $\frac{1}{c}$ against $\frac{1}{T}$
C c against $\frac{1}{T}$
D c against $c \times T$

17 Which one of the following properties of a catalyst can never be changed during the reaction which it catalyzes?
A chemical composition
B density
C mass
D physical state
E surface area

Classification

Questions 18–20

Figure 20.2 shows a graph of the volume of hydrogen produced against time during a rate of reaction experiment between zinc and dilute hydrochloric acid.

At which point on the graph

18 is the reaction complete? **19** is the reaction fastest? **20** are half the reactants used up?

Structured questions

21 An experiment was carried out using marble chips and hydrochloric acid to investigate the effects of particle size on the rate of reaction.

$$CaCO_3(s) + 2HCl(aq) \rightarrow CaCl_2(aq) + CO_2(g) + H_2O(l)$$

A large marble chip (mass 0.4 g) was placed in a conical flask and the flask placed on top of a pan balance. 25 cm^3 of dilute hydrochloric acid was added to the flask and a plug of cotton wool placed in the neck of the flask. The reading on the balance was noted at intervals.

The results obtained were

Time (minutes)	0	2	4	6	8	10	11	12
Total loss of mass (g)	0	2.2	2.9	3.3	3.6	3.7	3.7	3.7

(*a*) What is the chemical name for marble?
(*b*) Why is there a loss of mass during the experiment?
(*c*) Why was a plug of cotton wool placed in the neck of the flask?
(*d*) Plot a graph of the total loss of mass (on *y* axis) against time (on *x* axis).
(*e*) When was the reaction fastest?
(*f*) After how many minutes was the reaction completed?
(*g*) After 12 minutes a small piece of marble remained in the flask. What can be concluded about the amounts of marble and hydrochloric acid used?

The experiment was repeated using 0.4 g of powdered marble and a fresh 25 cm^3 sample of hydrochloric acid.

(*h*) What practical difficulty would be encountered in doing this? How could this difficulty be overcome?
(*i*) On the graph already drawn sketch the graph that would be obtained.

21 Reversible reactions and equilibrium

Simple multiple choice

1 An equilibrium is established with a reversible reaction when the rate of the
A forward reaction is greater than the rate of the reverse reaction.
B reverse reaction is greater than the rate of the forward reaction.
C forward reaction and the rate of the reverse reaction are zero.
D forward reaction is equal to the rate of the reverse reaction.

2 Calcium oxide can be made by heating calcium carbonate. In a closed system the following equilibrium is established.

$CaCO_3(s) \rightleftharpoons CaO(s) + CO_2(g)$

To ensure a good yield of calcium oxide
A the carbon dioxide must escape from the system.
B additional carbon dioxide must be added.
C the calcium carbonate must be crushed into small pieces.
D the system must be closed to prevent the escape of chemicals.
E the temperature must remain constant.

3 When iron and steam are heated together in a closed container the following equilibrium is established.

$3Fe(s) + 4H_2O(g) \rightleftharpoons Fe_3O_4(s) + 4H_2(g)$

The yield of hydrogen can be increased by
A adding a catalyst.
B allowing hydrogen to escape.
C adding some iron oxide.
D cooling down the system.
E using finely divided iron.

Questions 4–6
Nitrogen and hydrogen combine together to produce ammonia.

$N_2(g) + 3H_2(g) \rightleftharpoons 2NH_3(g) + \text{heat}$

4 What is the effect of a catalyst on the rate of establishing equilibrium and the yield of ammonia?

	Rate	*Yield*
A	increased	decreased
B	increased	unchanged
C	increased	increased
D	decreased	decreased
E	decreased	increased

5 What is the effect of increasing the temperature on the rate of establishing equilibrium and the yield of ammonia?

	Rate	*Yield*
A	increased	decreased
B	increased	unchanged
C	increased	increased
D	decreased	decreased
E	decreased	increased

6 In order to get the maximum conversion to ammonia it is necessary to have

A low temperature, high pressure, catalyst.
B low temperature, high pressure, no catalyst.
C high temperature, high pressure, catalyst.
D high temperature, high pressure, no catalyst.
E high temperature, low pressure, catalyst.

7 $Ag^+(aq) + Fe^{2+}(aq) \rightleftharpoons Ag(s) + Fe^{3+}(aq)$
Which one of the following would cause the equilibrium to move to the left?

A warm the solution.
B Add iron(II) sulphate solution.
C Add silver nitrate solution.
D Stir the solution.
E Add dilute hydrochloric acid to precipitate silver chloride.

8 $Fe^{3+}(aq) + 3CNS^-(aq) \rightleftharpoons Fe(CNS)_3(aq)$
pale yellow colourless blood red
What colour change would be observed if iron(III) chloride solution is added to an equilibrium mixture containing these ions?
The solution

A turns to a deeper red colour.
B turns colourless.
C turns to a pale yellow colour.
D contains a coloured precipitate.

Classification

Questions 9–13

A $N_2(g) + O_2(g) \rightleftharpoons 2NO(g)$
B $2HI(g) \rightleftharpoons H_2(g) + I_2(g)$
C $2SO_2(g) + O_2(g) \rightleftharpoons 2SO_3(g)$
D $PbCl_2(s) + aq \rightleftharpoons Pb^{2+}(aq) + 2Cl^-(aq)$
E $(NH_4)_2CO_3(s) \rightleftharpoons 2NH_3(g) + H_2O(l) + CO_2(g)$

In which one of A–E above

9 does an increase in pressure move the equilibrium to the right?

10 would a purple vapour be observed?

11 does a solid dissociate into ions in solution?

12 does a reaction take place which also occurs during a thunderstorm?

13 does a reaction occur which is a vital step in the Contact Process?

Multiple completion

Type 1 (See instructions on page 1.)

14 Which of the following statements about a system in equilibrium is (are) correct?
(*i*) All of the reactants are used up. (*ii*) All reactions have stopped.
(*iii*) Addition of more reactants would produce more products.

Type 2 (See instructions on page 1.)

15 $CH_3COOH(l) + C_2H_5OH(l) \rightleftharpoons CH_3COOC_2H_5(l) + H_2O(l)$
Which of the following statements is (are) true for the above equilibrium?
At equilibrium
(*i*) the rate of the forward reaction is equal to the rate of the reverse reaction.
(*ii*) the concentrations of all reactants and products are the same.
(*iii*) the concentrations of all reactants and products are independent of the external pressure.
(*iv*) the yield of ester could be significantly increased by adding a catalyst.

Assertion-reason

(See instructions on pages 1–2.)

16 When a system is in equilibrium the concentrations of all products and reactants remain unchanged. — In equilibrium, the rate of the forward reaction is equal to the rate of the reverse reaction.

Structured questions

17 The following equation summarizes the reaction which takes place between iron and steam.
$3Fe(s) + 4H_2O(g) \rightleftharpoons Fe_3O_4(s) + 4H_2(g)$ $H = -153$ kJ (heat given out)
(*a*) Explain briefly how samples of iron(II) diiron(III) oxide and hydrogen could be prepared from iron.
(*b*) Explain briefly how iron and water can be prepared from iron(II) diiron(III) oxide and hydrogen.
(*c*) When iron and steam are heated together at 400°C in a sealed container it is impossible to convert all of the reactants into products. Despite prolonged heating the concentrations of reactants and products remain unchanged.
(*i*) Explain why complete conversion is impossible.
(*ii*) What name is given to the situation existing in the container when no further reaction is apparently taking place?
(*iii*) What changes would occur to the system if the temperature was raised to 500°C?

22 Sulphur

Simple multiple choice

1 The process usually used for extracting sulphur from underground deposits involves
A decomposition.
B superheated water.
C reduction.
D distillation.
E electrolysis.

2 Sulphur is obtained from underground deposits by the Frasch process. Coal is not obtained by a similar process because coal
A is at greater depths than sulphur.
B does not react with superheated water.
C is not required in a pure form.
D would react with hot compressed air.
E is not melted by superheated water.

Classification

Questions 3–9

A α-sulphur B β-sulphur C Plastic sulphur D Amorphous sulphur

Which one of the above forms of sulphur is

3 formed when dilute hydrochloric acid is added to sodium thiosulphate solution?

4 formed initially when molten sulphur is cooled slowly?

5 formed initially when molten sulphur is cooled rapidly by pouring molten sulphur into cold water?

6 composed of long chains of sulphur atoms?

7 formed when sulphur is deposited from a solution following the evaporation of a volatile solvent at room temperature?

8 composed of octahedral crystals?

9 able to pass through a filter paper unchanged?

Questions 10–12

A solid B runny liquid C viscous liquid D gas

In which of the forms A–D above would sulphur exist at each temperature in the questions **10–12** below.

10 400°C

11 100°C

12 450°C

Assertion-reason

(See instructions on pages 1–2.)

The densities of α- and β-sulphur are 2.05 and 1.96 g per cm^3 respectively.

13 The volume of one mole of sulphur increases when it changes from α to β.	The sulphur atoms are more closely packed in α-sulphur than in β-sulphur.

23 Sulphides and hydrogen sulphide

Simple multiple choice

1 The gas which turns filter paper soaked in lead(II) nitrate solution black is
A sulphur dioxide.
B hydrogen sulphide.
C hydrogen.
D sulphur trioxide.
E carbon dioxide.

2 Which one of the following gases is a colourless gas with a pungent smell?
A nitrogen dioxide
B chlorine
C carbon dioxide
D hydrogen sulphide
E carbon monoxide

3 Which one of the following solutions reacts with hydrogen sulphide to form a precipitate of sulphur and a coloured solution?
A copper(II) sulphate solution
B lead(II) nitrate solution
C acidified potassium manganate(VII) solution
D chlorine water
E acidified potassium dichromate(VI) solution

4 Dilute hydrochloric acid is added to a mixture of finely powdered iron filings and sulphur. Which one of the following describes the change that occurs?
A Hydrogen gas is evolved.
B Hydrogen sulphide gas is evolved.
C The mixture dissolves leaving no residue.
D No change occurs.

5 Hydrogen sulphide and sulphur dioxide react as follows

$2H_2S(g) + SO_2(g) \rightarrow 2H_2O(l) + 3S(s)$

During the reaction

A hydrogen sulphide is oxidized and sulphur dioxide is reduced.
B hydrogen sulphide is reduced and sulphur dioxide is oxidized.
C hydrogen sulphide is oxidized and sulphur dioxide is oxidized.
D hydrogen sulphide is reduced and sulphur dioxide is reduced.

6 The salts obtained from hydrogen sulphide by replacing hydrogen ions are called
A sulphates. B sulphites. C thiosulphates. D sulphides.

7 Which is the correct equation for the reaction between iron(II) sulphide and dilute hydrochloric acid?

A $FeS_2(s) + 2HCl(aq) \rightarrow FeCl_2(aq) + H_2S(g) + S(s)$
B $Fe_2S_3(s) + 4HCl(aq) \rightarrow 2FeCl_2(aq) + 2H_2S(g) + S(s)$
C $FeS(s) + 2HCl(aq) \rightarrow FeCl_2(aq) + H_2S(g)$
D $2FeS(s) + 6HCl(aq) \rightarrow 2FeCl_3(aq) + 2H_2S(g) + S(s)$
E $Fe_2S_3(s) + 6HCl(aq) \rightarrow 2FeCl_3(aq) + 3H_2S(g)$

8 Which one of the following does **not** form a precipitate with hydrogen sulphide?
A lead(II) ethanoate
B sodium hydroxide solution
C chlorine water
D copper(II) sulphate solution
E an aqueous solution of sulphur dioxide

9 Hydrogen sulphide prepared from iron(II) sulphide and dilute hydrochloric acid contains considerable impurities. The major impurity is
A hydrogen chloride fumes.
B sulphur dioxide.
C hydrogen.
D sulphur trioxide.

10 Hydrogen sulphide can be prepared in a purer form by heating antimony(III) sulphide with concentrated hydrochloric acid. The equation for the reaction is

A $SbS(s) + 2HCl(aq) \rightarrow SbCl_2(aq) + H_2S(g)$.
B $SbS_2(s) + 2HCl(aq) \rightarrow SbCl_2(aq) + H_2S(g) + S(s)$.
C $2SbS(s) + 6HCl(aq) \rightarrow 2SbCl_3(aq) + 2H_2S(g) + S(s)$.
D $Sb_2S_3(s) + 6HCl(aq) \rightarrow 2SbCl_3(aq) + 3H_2S(g)$.
E $Sb_2S_3(s) + 4HCl(aq) \rightarrow 2SbCl_2(aq) + 2H_2S(g) + S(s)$.

Multiple completion

Type 2 (See instructions on page 1.)

11 The products from burning hydrogen sulphide in air may include
(*i*) water. (*ii*) sulphur. (*iii*) sulphur dioxide. (*iv*) sulphur trioxide.

Assertion-reason

(See instructions on pages 1–2.)

12 Hydrogen sulphide is collected over cold water.	Hydrogen sulphide is more soluble in hot water than in cold.
13 Lead-based paints blacken readily in industrial areas.	Hydrogen sulphide reacts with lead compounds to form black lead(II) sulphide.
14 When dilute hydrochloric acid is added to iron(II) sulphide, hydrogen sulphide, containing some hydrogen, is produced.	Iron(II) sulphide usually contains some unreacted iron.

24 Oxides of sulphur

Simple multiple choice

1 Which one of the following elements forms both an acidic oxide and an acidic hydride?
A sodium B aluminium C carbon D sulphur E nitrogen

2 Which one of the following oxides dissolves in water to give a solution with a pH less than 7?
A carbon monoxide
B sulphur dioxide
C aluminium oxide
D magnesium oxide
E sodium oxide

3 An element X forms an oxide XO_3 which reacts with water to form a strongly acidic solution. In which group of the Periodic Table is X placed?
A II B III C V D VI E VII

4 When a substance, in contact with air, is strongly heated, sulphur dioxide is produced. **On the basis of this information only,** it can be concluded that the substance must

A be sulphur.
B contain sulphur.
C be a sulphide.
D be a sulphate.
E be a sulphite.

5 Which one of the following statements about the usual laboratory preparation of sulphur dioxide is **incorrect?** The gas is

A collected by downward delivery.
B dried by passing through concentrated sulphuric acid.
C produced when the reactants are heated.
D produced when copper turnings and dilute sulphuric acid react together.

6 When sulphur trioxide bubbles through a solution of barium chloride, acidified with dilute hydrochloric acid, a white precipitate is formed.
The white precipitate is

A barium sulphate.
B barium hydroxide.
C barium sulphide.
D barium sulphite.
E barium chloride.

7 When sulphur dioxide is bubbled through a solution of sodium hydroxide a colourless solution is formed. Which one of the following compounds is present in the solution?

A sodium sulphate
B sodium sulphite
C sodium sulphide
D sodium hydrogensulphate

8 Both anhydrous iron(II) sulphate $FeSO_4$ and iron(III) sulphate $Fe_2(SO_4)_3$ decompose on heating to produce iron(III) oxide Fe_2O_3 and oxides of sulphur. Which one of the following statements is true?

A Both sulphates liberate only sulphur dioxide.
B Both sulphates liberate only sulphur trioxide.
C Iron(II) sulphate liberates sulphur dioxide only; iron(III) sulphate liberates sulphur trioxide only.
D Iron(II) sulphate liberates sulphur dioxide only; iron(III) sulphate liberates sulphur dioxide and sulphur trioxide.
E Iron(II) sulphate liberates sulphur dioxide and sulphur trioxide; iron(III) sulphate liberates sulphur trioxide only.

Questions 9 and 10 refer to the pieces of apparatus shown in Fig. 24.1.

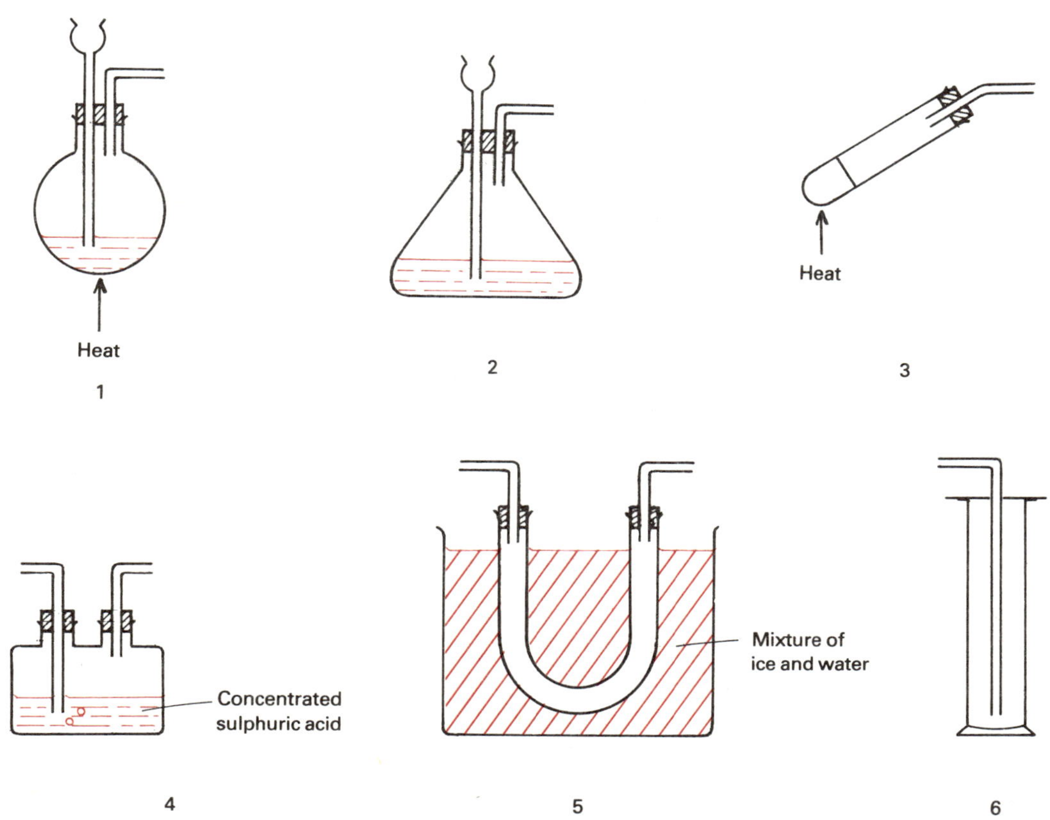

Fig. 24.1 **(continued on page 48)**

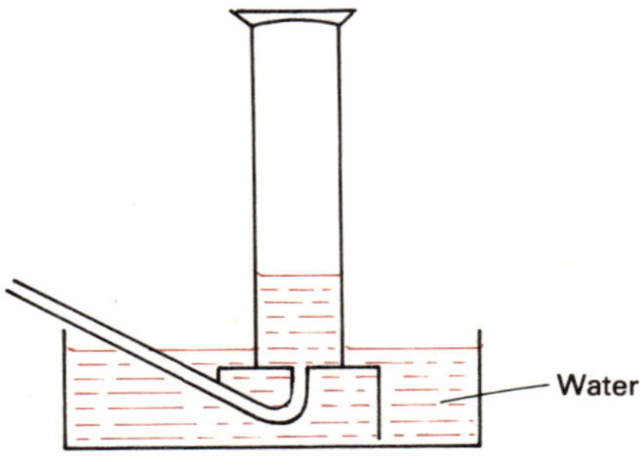

7

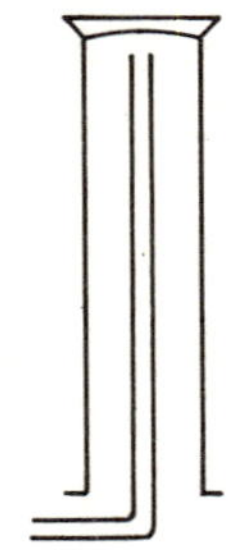

8

These pieces of apparatus can be connected together with rubber tubing.

9 Which of the combinations of apparatus in Fig. 24.1 (page 47 and above) would be most suitable for the preparation of a sample of dry sulphur dioxide from the reaction between sodium sulphite and dilute hydrochloric acid?
A 2, 7 B 2, 8 C 1, 4, 6 D 1, 4, 7 E 1, 4, 8

10 When some crystals of X are heated, sulphur dioxide, sulphur trioxide and water vapour are produced. Which of the combinations of apparatus in Fig. 24.1 would be most suitable to obtain some separate samples of sulphur dioxide and sulphur trioxide from crystals of X?
A 3, 4, 5, 6 B 3, 4, 5, 7 C 3, 4, 7 D 3, 5, 7 E 3, 5, 6

Multiple completion

Type 1 (See instructions on page 1.)

11 Sulphur dioxide can be prepared in the laboratory by
(*i*) burning sulphur. (*ii*) heating copper with concentrated sulphuric acid.
(*iii*) the action of dilute hydrochloric acid on a sulphate.

12 Sulphur trioxide can be prepared by
(*i*) burning sulphur. (*ii*) the action of acid on a sulphate.
(*iii*) the oxidation of sulphur dioxide with a catalyst.

13 Sulphur trioxide
(*i*) turns blue litmus red. (*ii*) forms steamy fumes in contact with moist air.
(*iii*) forms needle-shaped crystals on cooling.

Type 2 (See instructions on page 1.)

14 In which one of the following reactions is sulphur dioxide acting as an oxidizing agent?
(*i*) $H_2O(l) + SO_2(g) \rightleftharpoons H_2SO_3(aq)$
(*ii*) $2Mg(s) + SO_2(g) \rightarrow 2MgO(s) + S(s)$
(*iii*) $SO_2(aq) + 2H_2O(l) + Cl_2(g) \rightarrow H_2SO_4(aq) + 2HCl(aq)$
(*iv*) $2H_2S(g) + SO_2(g) \rightarrow 3S(s) + 2H_2O(l)$

Assertion-reason

(See instructions on pages 1–2.)

	Assertion	Reason
15	Sulphur dioxide, like carbon dioxide, turns limewater milky.	Both calcium carbonate and calcium sulphite are insoluble.
16	Both hydrogen sulphide and sulphur dioxide turn potassium dichromate(VI) solution from orange to green.	During the reactions of hydrogen sulphide and sulphur dioxide with potassium dichromate(VI), a precipitate of sulphur is formed with hydrogen sulphide but not with sulphur dioxide.

25 Sulphuric acid and sulphates

Simple multiple choice

1 Which one of the following pairs of substances could be separated by the addition of dilute sulphuric acid followed by filtration?
A copper and copper(II) oxide B magnesium and magnesium oxide
C sodium carbonate and copper(II) carbonate
D lead(II) nitrate and lead(II) sulphate
E barium chloride and barium carbonate

2 Which one of the following reagents would distinguish the presence of sulphuric acid in a mixture of sulphuric and hydrochloric acids?

A Universal indicator
B silver nitrate solution
C barium chloride solution
D litmus solution
E sodium hydroxide solution

3 A few drops of concentrated sulphuric acid are added to some water in a test tube and then a drop of Universal indicator is added. Which one of the following best describes the observations that are made?

The solution is

A above room temperature and purple in colour.
B above room temperature and red in colour.
C at room temperature and green in colour.
D at room temperature and red in colour.
E below room temperature and red in colour.

4 In which one of the following reactions involving sulphuric acid is sulphuric acid acting as an oxidizing agent?

A $CuO(s) + H_2SO_4(aq) \rightarrow CuSO_4(aq) + H_2O(l)$
B $NaOH(aq) + H_2SO_4(aq) \rightarrow NaHSO_4(aq) + H_2O(l)$
C $Na_2CO_3(s) + H_2SO_4(aq) \rightarrow Na_2SO_4(aq) + CO_2(g) + H_2O(l)$
D $Cu(s) + 2H_2SO_4(l) \rightarrow CuSO_4(aq) + 2H_2O(l) + SO_2(g)$
E $Cu(OH)_2(s) + H_2SO_4(aq) \rightarrow CuSO_4(aq) + 2H_2O(l)$

5 Concentrated sulphuric acid reacts with potassium chloride to produce

A chlorine.
B hydrogen.
C sulphur dioxide.
D hydrogen chloride.
E sulphur trioxide.

6 When concentrated sulphuric acid is added to blue copper(II) sulphate-5-water crystals the crystals turn white. The acid is acting as

A a dehydrating agent.
B an oxidizing agent.
C a reducing agent.
D a catalyst.

7 The sulphur trioxide produced in the Contact process is

A dissolved in water.
B dissolved in dilute sulphuric acid.
C dissolved in concentrated sulphuric acid.
D liquefied by applying pressure.
E cooled to form needle-shaped crystals.

8 Ethene can be absorbed in concentrated sulphuric acid and then the solution diluted with water. The product is ethanol. This reaction is an example of

A dehydration.
B neutralization.
C esterification.
D hydration.
E dissolving.

9 When methanoic acid (formic acid) is warmed with concentrated sulphuric acid, the gaseous product is

A carbon dioxide.
B sulphur dioxide.
C water.
D hydrogen sulphide.
E carbon monoxide.

10 Which one of the following sulphates is insoluble in water?

A sodium hydrogensulphate
B potassium sulphate
C lead(II) sulphate
D zinc sulphate
E iron(II) sulphate

11 When concentrated sulphuric acid is carefully poured down the side of a test tube containing water, two layers are formed with sulphuric acid in the lower layer. This is because sulphuric acid

A has a lower boiling point than water.
B does not react with water.
C does not dissolve in water.
D has a higher density than water.
E has a lower density than water.

Classification

Questions 12–16

Sulphuric acid can

A act as an oxidizing agent.
B be used to make insoluble sulphates.
C act as an acid.
D act as a dehydrating agent.
E act as a catalyst.

For each of the reactions in questions **12–16** choose A–E above which describes the reaction.

12 $Pb(NO_3)_2(aq) + H_2SO_4(aq) \rightarrow PbSO_4(s) + 2HNO_3(aq)$

13 $C_{12}H_{22}O_{11}(s) \rightarrow 12C(s) + 11H_2O(g)$

14 $2NaOH(aq) + H_2SO_4(aq) \rightarrow Na_2SO_4(aq) + 2H_2O(l)$

15 $C_2H_5OH(g) \rightarrow C_2H_4(g) + H_2O(g)$

16 $2FeSO_4(aq) + 2H_2SO_4(l) \rightarrow Fe_2(SO_4)_3(aq) + 2H_2O(l) + SO_2(g)$

Multiple completion

Type 1 (See instructions on page 1.)

17 Which of the following substances are dehydrated by concentrated sulphuric acid?
(*i*) ethanedioic (oxalic) acid (*ii*) glucose (*iii*) ethanol

18 Excess concentrated sulphuric acid reacts with
(*i*) potassium nitrate. (*ii*) copper. (*iii*) platinum.

Type 2 (See instructions on page 1.)

19 Sulphur trioxide is produced by the oxidation of sulphur dioxide. Which of the following methods could provide a source of sulphur dioxide for the process?
(*i*) burning of sulphur (*ii*) reaction of copper with concentrated sulphuric acid
(*iii*) roasting of metal sulphides in air (*iv*) action of dilute acids on metal sulphides

Assertion-reason

(See instructions on pages 1–2.)

20 When diluting concentrated sulphuric acid with water it is essential to add the acid to the water.

The dilution of concentrated sulphuric acid is an exothermic process.

26 Nitrogen

Simple multiple choice

1 A colourless, odourless gas extinguishes a lighted splint and does not turn limewater milky. This gas could be
A hydrogen. C nitrogen. E ammonia.
B carbon dioxide. D carbon monoxide.

2 Which one of the following pairs of compounds produces nitrogen on heating?
A Ammonium chloride and sodium nitrate B Ammonium sulphate and sodium nitrite
C Ammonium nitrate and sodium chloride D Ammonium sulphate and sodium hydroxide
E Ammonium chloride and sodium hydroxide

Questions 3 and 4 concern the preparation of a sample of nitrogen from air using the apparatus shown in Fig. 26.1.

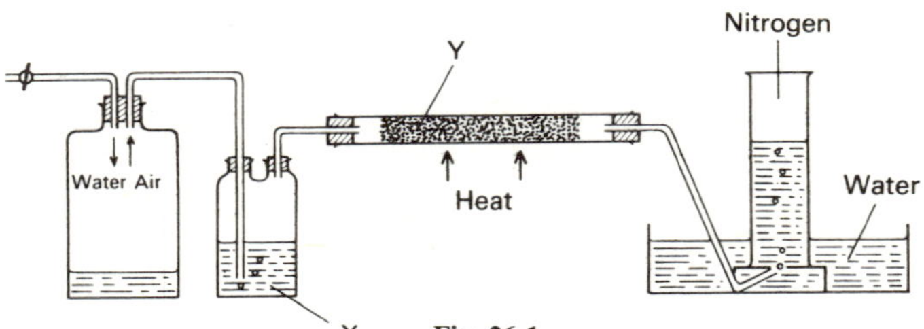

Fig. 26.1

3 The two substances labelled X and Y could be

	X	Y
A	sodium hydroxide solution	copper(II) oxide
B	concentrated sulphuric acid	copper(II) oxide
C	concentrated sulphuric acid	copper
D	sodium hydroxide solution	copper

4 Apart from water vapour, the principal impurity in the nitrogen collected is
A carbon dioxide. B oxygen. C hydrogen. D argon. E krypton.

5 Which one of the following does not supply nitrogen to the soil?
A bacteria B lime C plant decay D animal decay

6 Nitrogen is removed from the soil in the form of nitrates by
A absorption through the roots of plants.
B nitrifying bacteria. C lightning.
D the manufacture of nitric acid E the Haber process.

7 Element Z is a colourless gas at room temperature. Z reacts with magnesium on heating to form a white solid Mg_3Z_2. This solid reacts with water to form an alkaline gas ZH_3. These properties suggest that Z could be
A oxygen. B sulphur. C chlorine. D nitrogen. E phosphorus.

Multiple completion

Type 1 (See instructions on page 1.)

8 Nitrogen enters the soil by means of
(*i*) lightning and rain. (*ii*) plants with root nodules. (*iii*) man-made fertilizers.

9 Which elements combine directly with nitrogen to form compounds?
(*i*) lithium (*ii*) chlorine (*iii*) carbon

Type 2 (See instructions on page 1.)

10 Which of the following compounds evolve nitrogen on heating?
(*i*) ammonium nitrite
(*ii*) ammonium nitrate
(*iii*) ammonium dichromate(VI)
(*iv*) ammonium sulphate

27 Ammonia and ammonium compounds

Simple multiple choice

1 Ammonia gas is usually collected by upward delivery. The reason for using this method for collecting ammonia is that it is
A very soluble in water. B much less dense than air. C alkaline. D colourless.

2 When ammonia is burnt in oxygen the products are
A nitrogen and water.
B nitrogen monoxide and hydrogen.
C nitrogen dioxide and water.
D dinitrogen monoxide and hydrogen.
E dinitrogen monoxide and water.

3 When a food sample is heated with soda-lime, an alkaline gas is evolved. Which one of the following is likely to be in the food?
A vitamins B carbohydrates C proteins D fats E minerals

4 The drying agent usually used to dry ammonia is
A calcium carbonate.
B concentrated sulphuric acid.
C dilute sulphuric acid.
D calcium oxide.
E anhydrous calcium chloride.

5 A colourless gas is very soluble in water. The gas when passed over heated copper(II) oxide converts the copper(II) oxide to copper.
The gas could be
A hydrogen.
B hydrogen chloride.
C ammonia.
D carbon monoxide.
E carbon dioxide.

Questions 6–9

Methylamine CH_3NH_2 is an alkaline gas, more dense than air and soluble in water. It resembles ammonia but burns readily in air.

6 Methylamine can be prepared by heating $CH_3NH_3^+Cl^-$ with
A sodium hydroxide solution.
B sodium chloride solution.
C dilute hydrochloric acid.
D concentrated sulphuric acid.

7 Methylamine can be collected dry by
A collecting over water.
B passing through concentrated sulphuric acid and collecting by downward delivery.
C passing through concentrated sulphuric acid and collecting by upward delivery.
D passing over calcium oxide and collecting by downward delivery.
E passing over calcium oxide and collecting by upward delivery.

8 Methylamine and ammonia can be distinguished by adding
A Universal indicator.
B a lighted splint.
C litmus paper.
D limewater.
E a glowing splint.

9 A solution of methylamine in water is **alkaline** because of the presence of
A NH_4^+. B $CH_3NH_3^+$. C H^+. D OH^-.

10 The fountain experiment is used to demonstrate that ammonia is
A an alkaline gas.
B a colourless gas.
C less dense than air.
D a reactive gas.
E readily soluble in water.

Questions 11 and 12

When a piece of hot platinum wire is placed in a mixture of ammonia and air, the wire glows brighter and brown fumes are observed.

11 What is the purpose of the platinum?
The platinum
A reacts with the ammonia.
B reacts with the air.
C acts as a catalyst.
D acts as a source of heat.

12 The brown fumes produced are
A nitric acid. B nitrogen monoxide. C bromine. D nitrogen dioxide.

Classification

Questions 13–15
A $8NH_3(g) + 3Cl_2(g) \rightarrow N_2(g) + 6NH_4Cl(s)$
B $NH_3(g) + H_2O(l) \rightleftharpoons NH_4^+(aq) + OH^-(aq)$
C $4NH_3(g) + 3O_2(g) \rightarrow 2N_2(g) + 6H_2O(g)$
D $4NH_3(g) + 5O_2(g) \rightarrow 4NO(g) + 6H_2O(g)$
E $2NH_3(g) + 3CuO(s) \rightarrow 3Cu(s) + 3H_2O(g) + N_2(g)$

In which one of the reactions A–E above is

13 ammonia not being oxidized?

14 a catalyst essential?

15 ammonia burning in oxygen under ordinary laboratory conditions?

Multiple completion

Type 2 (See instructions on page 1.)

16 Which of the following pairs of substances will evolve **only** ammonia gas on heating?
(*i*) ammonium chloride and concentrated sulphuric acid
(*ii*) ammonium sulphate and sodium chloride
(*iii*) ammonium nitrate and sodium hydroxide
(*iv*) ammonium sulphate and calcium hydroxide

17 Which of the following statements about the Haber process are true?
(*i*) The volume of hydrogen used is three times the volume of nitrogen.
(*ii*) The catalyst is finely divided iron. (*iii*) Ammonia is removed by liquefying.
(*iv*) The conversion of nitrogen and hydrogen to ammonia occurs best at very high temperatures.

Assertion-reason

(See instructions on pages 1–2.)

18 When 20 cm^3 of ammonia gas are decomposed, the volume of the resulting gases is 40 cm^3.	Two moles of ammonia molecules produce one mole of nitrogen molecules and three moles of hydrogen molecules.
19 When ammonia gas is passed over heated lead(II) oxide a residue of lead is produced.	Ammonia is frequently used as an oxidizing agent.
20 The solubilities of ammonium salts in water closely resemble the solubilities of alkali metal salts.	Both alkali metal and ammonium ions have a single positive charge.

28 Nitric acid and nitrates

Simple multiple choice

1 Which one of the following pairs of substances produces nitric acid vapour on heating?
A potassium nitrite and dilute sulphuric acid
B potassium nitrite and concentrated sulphuric acid
C potassium nitrate and sodium hydroxide solution
D potassium nitrate and dilute sulphuric acid
E potassium nitrate and concentrated sulphuric acid

Questions 2 and 3
Concentrated nitric acid is frequently yellow in colour.

2 This yellow colour is due to
A a shortage of air in the acid. B the presence of air in the acid.
C the presence of iron(III) impurities. D the presence of nitrogen dioxide impurities.

3 The yellow colour can be removed by
A passing air through the solution. C adding water. E heating.
B distillation. D filtering.

4 The action of heat on lead(II) nitrate produces
A lead(II) nitrite and oxygen.
B lead(II) oxide, nitrogen dioxide and oxygen.
C lead(IV) oxide and nitrogen dioxide. D lead, nitrogen dioxide and oxygen.
E lead(IV) oxide, nitrogen monoxide and oxygen.

5 Which one of the following produces an amphoteric oxide on heating?
A copper(II) nitrate C zinc nitrate E calcium nitrate
B sodium nitrate D magnesium nitrate

6 Which one of the following metal nitrates does **not** form a metal oxide on heating?
A lead(II) nitrate C calcium nitrate E zinc nitrate
B silver nitrate D copper(II) nitrate

7 Which one of the following equations is correct?
A $MgOH(s) + HNO_3(aq) \rightarrow MgNO_3(aq) + H_2O(l)$
B $MgOH_2(s) + HNO_3(aq) \rightarrow MgHNO_3(aq) + H_2O(l)$
C $Mg(OH)_2(s) + 2HNO_3(aq) \rightarrow MgNO_3(aq) + 2H_2O(l) + NO_2(g)$
D $Mg(OH)_2(s) + 2HNO_3(aq) \rightarrow Mg(NO_3)_2(aq) + 2H_2O(l)$

Multiple completion

Type 1 (See instructions on page 1.)

8 When lead(II) nitrate crystals are heated
(*i*) nitrogen dioxide is evolved.
(*ii*) oxygen is evolved. (*iii*) a metallic residue remains.

9 In order to produce some hydrogen from the reaction between magnesium and nitric acid, it is necessary to
(*i*) use very dilute acid. (*ii*) heat strongly.
(*iii*) clean the magnesium thoroughly with sandpaper.

10 Nitric acid is a monobasic acid. It therefore
(*i*) forms aqueous solutions with a pH greater than 7.
(*ii*) has no oxidizing or reducing properties.
(*iii*) does not form acid salts.

29 Oxides of nitrogen

Simple multiple choice

1 Which one of the following gases is colourless but turns brown on contact with air?
A nitrogen B nitrogen monoxide C dinitrogen oxide D nitrogen dioxide

2 Which one of the following gases relights a glowing splint?
A nitrogen B nitrogen monoxide C nitrogen dioxide D dinitrogen monoxide

3 A solution containing sodium nitrate and sodium nitrite can be produced by adding which of the following to a solution of sodium hydroxide in water?
A nitric acid B dinitrogen monoxide C nitrogen dioxide D nitrogen monoxide

4 In which one of the following reactions is nitrogen dioxide **not** a product of the reaction?
A reaction of copper and concentrated nitric acid
B action of heat on copper(II) nitrate crystals
C reaction of nitrogen monoxide with oxygen
D thermal decomposition of nitric acid
E action of heat on ammonium nitrate

5 Arrange the three oxides of nitrogen – dinitrogen monoxide, nitrogen monoxide, nitrogen dioxide – in order of solubility with the most soluble first.
A dinitrogen oxide, nitrogen monoxide, nitrogen dioxide
B nitrogen dioxide, dinitrogen oxide, nitrogen monoxide
C nitrogen monoxide, nitrogen dioxide, dinitrogen oxide
D nitrogen dioxide, nitrogen monoxide, dinitrogen oxide

6 Refer to the apparatus in Fig. 24.1 on pages 47–8. Which of the following combinations of apparatus can be used to prepare a sample of nitrogen monoxide from copper and dilute nitric acid?
A 1, 7 B 2, 7 C 1, 6 D 1, 8 E 2, 8

7 Refer to the apparatus in Fig. 24.1 on pages 47–8. Which of the following combinations of apparatus can be used to collect samples of nitrogen tetroxide and oxygen from lead(II) nitrate crystals?
A 3, 5, 7 B 3, 5, 6 C 3, 5, 8 D 1, 5, 8 E 1, 5, 7

8 Which one of the following gases contains the greatest percentage of nitrogen by mass? ($A_r(N) = 14, A_r(H) = 1, A_r(O) = 16$)
A ammonia, NH_3 C dinitrogen oxide, N_2O
B nitrogen monoxide, NO D nitrogen dioxide, NO_2

9 An oxide of nitrogen and an oxide of carbon have the same density.
(A_r(C) = 12, A_r(N) = 14, A_r(O) = 16)
They are
A carbon monoxide and dinitrogen oxide.
B carbon dioxide and dinitrogen oxide.
C carbon dioxide and nitrogen dioxide.
D carbon monoxide and nitrogen monoxide.

10 Which one of the following gases is produced by the action of heat on a mixture of ammonium sulphate and sodium nitrate?
A nitrogen B nitrogen monoxide C nitrogen dioxide D dinitrogen oxide

30 Agricultural chemistry

Simple multiple choice

1 An NPK fertilizer contains
A nitrogen, phosphorus, potassium.
B nitrogen, phosphorus, krypton.
C nitrogen, potassium, krypton.
D nitrogen, potassium, sodium.
E sodium, phosphorus, potassium.

2 Which one of the following pairs of substances when dissolved in water would produce a complete fertilizer?
A ammonium sulphate and potassium nitrate
B calcium phosphate and ammonium nitrate
C ammonium phosphate and potassium sulphate
D potassium nitrate and calcium hydroxide
E potassium chloride and calcium nitrate

3 Iron and boron are required in small quantities by healthy plants.
These elements are called
A alkaline earth metals.
B transition metals.
C rare earth metals.
D trace elements.

4 Which one of the following fertilizers should **not** be applied to the soil with lime?
A potassium nitrate
B ammonium sulphate
C potassium sulphate
D calcium nitrate

5 One of the chief constituents of animal bones is
A nitrogen. B sodium. C phosphorus. D potassium.

6 Which one of the following fertilizers contains the highest percentage of nitrogen?
(A_r(H) = 1, A_r(C) = 12, A_r(N) = 14, A_r(O) = 16, A_r(S) = 32)
A ammonium sulphate, $(NH_4)_2SO_4$
B Urea, $CO(NH_2)_2$
C ammonium nitrate, NH_4NO_3
D ammonium carbonate, $(NH_4)_2CO_3$

Multiple completion

Type 1 (See instructions on page 1.)

7 Which of the following elements return to the soil on the burning of the stubble after harvesting a field of barley?
(*i*) potassium (*ii*) nitrogen (*iii*) phosphorus

Assertion-reason

(See instructions on pages 1–2.)

8 Calcium superphosphate is a more effective fertilizer than calcium phosphate. — Calcium superphosphate is more soluble in water than calcium phosphate.

9 Bonemeal and slag are good sources of potassium. — Potassium is necessary for the development of flowers and seeds.

10 Urea contains a greater percentage of nitrogen than ammonium nitrate. — Urea is a better fertilizer for applying to grass in Spring to promote rapid growth.

Structured questions

This question uses information from units 8, 27, 28 and 29.

11 Fig. 30.1 shows a flow diagram for the artificial fixation of nitrogen in two solid fertilizers Y and Z.
(*a*) What is meant by the term 'fixation of nitrogen'?
(*b*) Identify the two elements A and B.
(*c*) Identify the catalyst C used in the oxidation of ammonia.
(*d*) Identify the two oxides of nitrogen D and E.
(*e*) Identify the acid F.
(*f*) Identify the two solid fertilizers Y and Z.
(*g*) Name the three processes I, II and III taking place.

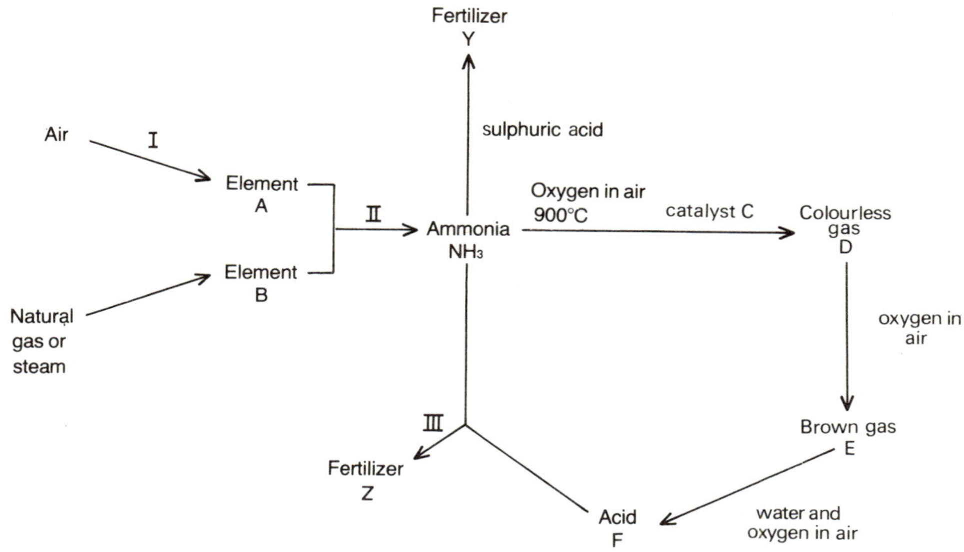

Fig. 30.1

31 Phosphorus

Simple multiple choice

1 Sand is added to the furnace for the extraction of phosphorus in order to
A lower the melting point of the mixture in the furnace.
B increase the electrical conductivity of the mixture in the furnace.
C remove impurities by forming slag.
D reduce phosphorus(V) oxide to phosphorus.

2 the chemical name for slag is
A calcium phosphate. C calcium carbonate. E calcium silicide.
B calcium sulphate. D calcium silicate.

3 Which one of the following statements about white phosphorus is **incorrect?**
White phosphorus is
A stored under oil. B spontaneously inflammable in air.
C soluble in carbon disulphide and other organic solvents.
D formed when phosphorus vapour is condensed.

4 When white phosphorus is heated to 250°C in an inert atmosphere in the presence of iodine it
A burns. C remains unchanged.
B turns to red phosphorus. D forms phosphorus(III) iodide.

5 When red phosphorus is heated to 450°C in an atmosphere of nitrogen, the phosphorus
A melts. D sublimes.
B remains unchanged. E forms phosphorus nitride.
C burns.

Multiple completion

Type 2 (See instructions on page 1.)

6 The raw materials in the production of phosphorus are
(*i*) calcium phosphate. (*ii*) coke. (*iii*) sand. (*iv*) limestone.

7 Phosphorus and its compounds are used for
(*i*) incendiary bombs. (*ii*) fertilizers. (*iii*) making matches.
(*iv*) in the manufacture of good quality steels.

Assertion-reason

(See instructions on pages 1–2.)

8 White phosphorus has a low melting point and is readily soluble in organic solvents.	White phosphorus has a structure composed of P_4 molecules.
9 Phosphorus is much less reactive than nitrogen.	Nitrogen and phosphorus are in the same group in the Periodic Table.
10 P_4O_{10} is reduced to phosphorus by electrolysis.	Electrolysis is an efficient method of extracting reactive metals from their ores.

32 Hydrogen chloride and chlorine

Simple multiple choice

1 Which one of the following gases turns damp blue litmus red and then white?
A chlorine B hydrogen chloride C carbon dioxide D oxygen

2 Chlorine is prepared by the action of
A concentrated sulphuric acid on sodium chloride.
B hot, concentrated hydrochloric acid on sodium chloride.
C hot, concentrated hydrochloric acid on an oxidizing agent.
D hot, concentrated hydrochloric acid on a reducing agent.
E dry hydrogen chloride on a heated metal.

3 Which one of the following statements about the preparation of pure, dry chlorine is incorrect?
A The reaction mixture is heated. B Manganese(IV) oxide acts as a catalyst.
C The gas is passed through water. D The gas is dried with concentrated sulphuric acid.
E The gas is collected by downward delivery.

4 The gas evolved when concentrated sulphuric acid is added to a mixture of manganese(II) chloride and manganese(IV) oxide is
A oxygen. C chlorine. E sulphur trioxide
B hydrogen chloride. D sulphur dioxide.

5 Which one of the following would be used for bleaching a piece of coloured cloth?
A sodium chlorate(I) (hypochlorite) solution
B soapless detergent solution C tetrachloromethane
D sodium hydroxide solution E soap solution

6 On heating hydrated iron(III) chloride a pungent gas is evolved. This gas turns blue litmus red and forms dense white fumes with ammonia.
The gas evolved is
A hydrogen. C chlorine. E none of these gases.
B hydrogen chloride. D ammonia.

Questions 7–9
When bleaching powder is treated with dilute hydrochloric acid, a gas X is evolved which turns blue litmus red and then bleaches it.
When cobalt(II) chloride solution is added to bleaching powder, a colourless, odourless gas Y is evolved which has no effect on indicators and relights a glowing splint.

7 The gas X is
A chlorine. C oxygen. E sulphur dioxide.
B hydrogen chloride. D hydrogen.

8 The gas Y is
A chlorine. C oxygen. E dinitrogen monoxide.
B hydrogen chloride. D hydrogen

9 In the reaction producing Y, the cobalt(II) chloride acts as
A an oxidizing agent. B a reducing agent. C an acid. D a base. E a catalyst.

Multiple completion

Type 1 (See instructions on page 1.)

10 Which of the following compounds liberate chlorine from concentrated hydrochloric acid?
(*i*) lead(IV) oxide
(*ii*) potassium manganate(VII) (permanganate) (*iii*) lead(II) oxide

11 A concentrated solution of hydrogen chloride in water
(*i*) bleaches indicator paper. (*ii*) produces chlorine gas when heated with an oxidizing agent.
(*iii*) produces hydrogen with magnesium.

12 Chlorine can be produced in industry by the
(*i*) electrolysis of molten sodium chloride.
(*ii*) electrolysis of an aqueous solution of sodium chloride.
(*iii*) action of heat on metal chlorides.

Type 2 (See instructions on page 1.)

13 Hydrogen chloride
(*i*) turns damp blue litmus paper red.
(*ii*) forms dense white fumes with the gas escaping from ammonia solution.
(*iii*) forms steamy fumes when in contact with moist air. (*iv*) has a greenish yellow colour.

14 Which of the following are properties of a solution of hydrogen chloride (salt gas) in water? It forms

(*i*) hydrogen with magnesium. (*ii*) hydrogen with copper.
(*iii*) carbon dioxide with sodium carbonate. (*iv*) chlorine with copper(II) oxide.

Assertion-reason

(See instructions on pages 1–2.)

15 Dry hydrogen chloride can be used in the fountain experiment.	Hydrogen chloride is very soluble in water at room temperature.
16 When chlorine is bubbled through an aqueous solution of hydrogen sulphide a white precipitate of sulphur is produced.	Chlorine is a strong reducing agent.
17 A solution of dry hydrogen chloride in dry methylbenzene has a pH of 1.	Hydrogen chloride dissolved in water exists in the form of $H^+(aq)$ and $Cl^-(aq)$ ions.

Structured questions

18 Figure 32.1 shows the apparatus necessary for preparing 'salt gas' from salt.

(*a*) Name the liquid which produces 'salt gas' when added to salt.
(*b*) What is the liquid in the wash bottle labelled X?
(*c*) Why is the 'salt gas' passed through the liquid in the wash bottle?
(*d*) Why is the 'salt gas' collected by this method?

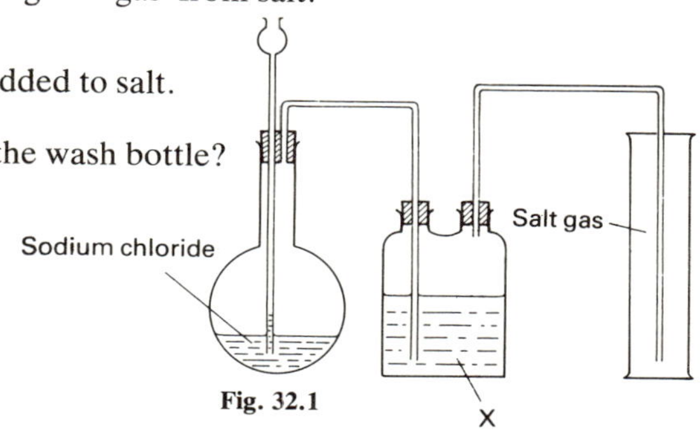

Fig. 32.1

(*e*) When 'salt gas' is passed over heated iron filings a reaction takes place producing a colourless gas which burns with a squeaky pop when a lighted splint is put into the gas.
(*i*) What is the gas produced when 'salt gas' reacts with iron?
(*ii*) Draw a diagram of the apparatus that could be used to prepare 'salt gas', pass it over heated iron filings and collect a sample of the gas produced free of 'salt gas'.
(*iii*) What can be concluded about the composition of 'salt gas' from the experiment with iron filings?

(*f*) What precautions should be taken when preparing a solution of 'salt gas' in water?
Experiments were carried out to compare the properties of solutions of dry 'salt gas' in water and dry methylbenzene. The results are summarized in the following table.

	Solution of 'salt gas' in water	*Solution of 'salt gas' in dry methylbenzene*
Add Universal indicator	pH 1	pH 7
Add calcium carbonate	Bubbles of colourless gas	No reaction
Add magnesium ribbon	Bubbles of colourless gas	No reaction
Electrical conductivity	Conductor	Non conductor

Table 32.1

(*g*) What can be concluded about each solution as a result of the tests with Universal indicator?
(*h*) Name the gas evolved when a solution of 'salt gas' in water is added to
(*i*) calcium carbonate, (*ii*) magnesium ribbon?
(*i*) What is the more common name for a solution of 'salt gas' in water?
(*j*) Explain the difference in the structure of 'salt gas' in water and methylbenzene?

33 Carbon, carbon monoxide and organic chemistry

Simple multiple choice

1 A colourless gas burns in oxygen with a blue flame to produce a single gaseous product. This new gas turns limewater milky. The original gas is

A carbon dioxide.
B carbon monoxide.
C hydrogen.
D methane.
E nitrogen.

2 Which one of the following processes is used to obtain petrol, paraffin and lubricating oils from crude oil?
A combustion
B fractional distillation
C cracking
D polymerization
E filtration

3 The gaseous product of the fermentation of glucose is
A oxygen.
B ethanol.
C carbon monoxide.
D carbon dioxide.
E hydrogen.

4 The products obtained following the complete combustion of petrol in excess air are
A carbon monoxide and water.
B carbon dioxide and hydrogen.
C carbon dioxide and water.
D carbon monoxide, carbon dioxide and water.
E carbon and water.

5 The chief danger of running a car engine in an enclosed garage is
A the lack of oxygen in the garage.
B the presence of petrol fumes which are very flammable.
C the formation of carbon dioxide.
D the build up of lead compounds in the garage.
E the formation of carbon monoxide.

6 Crude oil can be separated into different fractions because the fractions have different
A boiling points.
B melting points.
C densities.
D solubilities.
E colours.

7 The molecular formulae of the first three members of the alkene family are
A C_2H_2, C_2H_4, C_2H_6.
B CH_2, C_2H_4, C_3H_6.
C C_2H_4, C_3H_6, C_4H_8.
D C_2H_4, C_3H_8, C_4H_8.

8 Petrol in a car engine vaporizes and then burns. These changes are
A both physical changes.
B both chemical changes.
C a physical change followed by a chemical change.
D a chemical change followed by a physical change.

9 Starch in foods can be detected using
A iodine solution.
B Fehling's (or Benedict's) solution.
C saliva.
D litmus or Universal indicator.
E bromine water.

10 The adjacent equation represents the

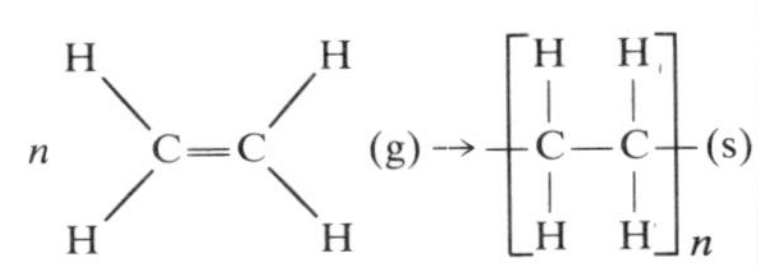

A oxidation of ethene.
B solidification of ethene.
C decomosition of ethene.
D melting of ethene.
E polymerization of ethene.

11 $CO(g) + PbO(s) \rightarrow CO_2(g) + Pb(s)$
In this reaction carbon monoxide acts as
A a catalyst.
B an acid.
C a base.
D an oxidizing agent.
E a reducing agent.

Questions 12 and 13

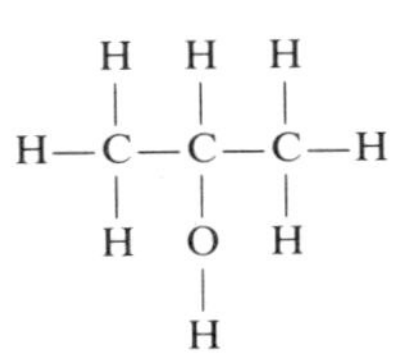

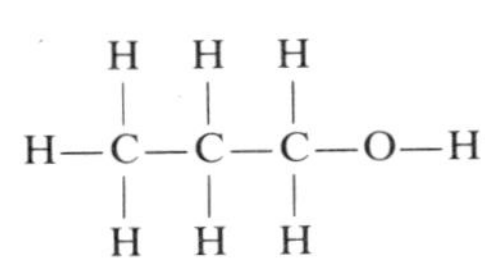

12 The two compounds above are
A allotropes.
B isotopes.
C polymorphs.
D isomers.
E polymers.

13 From their formulae alone it is possible to predict that they will have the same
A percentage composition by mass.
B melting points.
C boiling points.
D structural formulae.
E solubilities in water.

14 Which one of the following substances is a saturated hydrocarbon?
A glucose
B ethyne
C pentane
D octene
E propanol

15 Which one of the following compounds is produced when bromine is reacted with propene $CH_3CH{=}CH_2$?

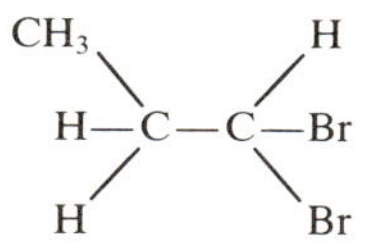

A

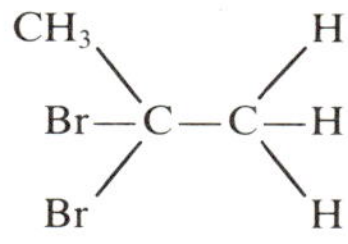

B

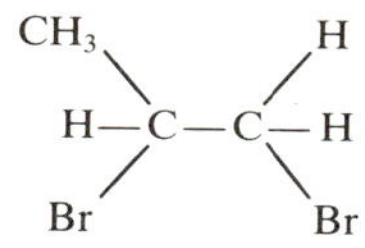

C

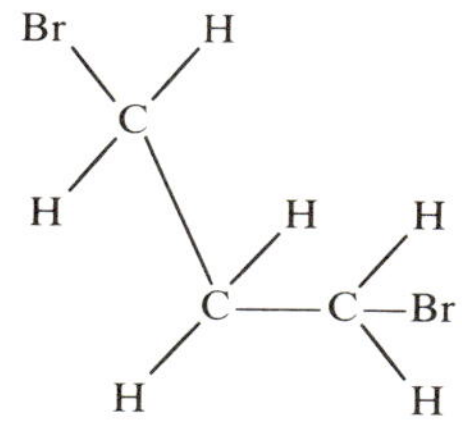

D

16 The carbon atoms in diamond and graphite differ in their
A mass.
B spatial arrangement.
C mass number.
D volume.
E colour.

17 Which one of the following would rapidly decolorize a solution of bromine dissolved in hexane?
A ethane B carbon monoxide C carbon dioxide D ethene E methane

18 A sample of glucose was mixed with copper(II) oxide and the mixture heated. The products of the reaction were carbon dioxide, water and copper. **From these results alone,** it can be concluded that glucose contains at least
A carbon and hydrogen. B carbon, hydrogen and oxygen.
C carbon, hydrogen, oxygen and copper.
D carbon and copper. E carbon, copper and oxygen.

19 Which one of the following compounds is formed when ethanol reacts with ethanoic acid in the presence of concentrated sulphuric acid?
A C_2H_4
B $CH_3CH_2COOCH_3$
C $CH_3CH_2COOCH_2CH_3$
D $CH_3CH_2CH_2COOH$
E $CH_3COOCH_2CH_3$

20 Which one of the following pairs of substances represents two isomers?
A diamond and graphite B carbon-12 and carbon-14

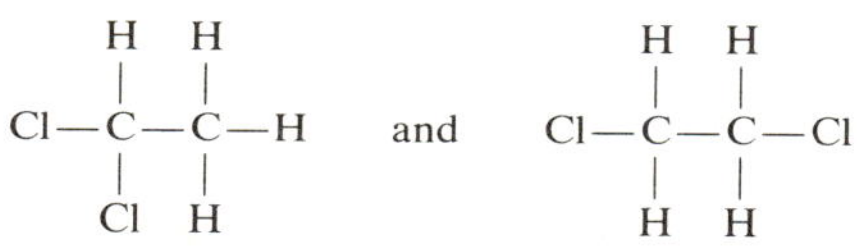

C

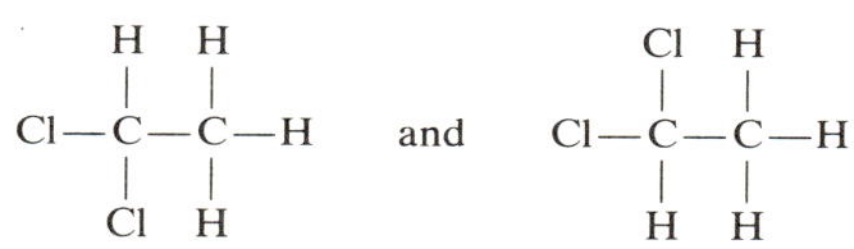

D

Questions 21–23
Figure 33.1 shows the apparatus required for producing ethene from liquid paraffin.

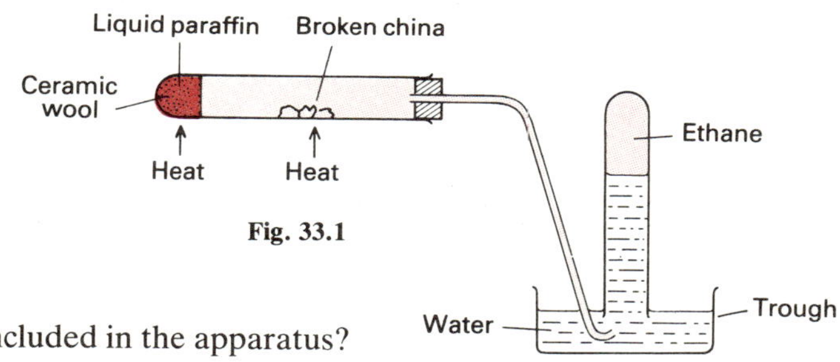

Fig. 33.1

21 For what purpose is ceramic wool included in the apparatus?
A to act as a catalyst
B to keep the heat in the apparatus
C to lower the boiling point of the liquid paraffin
D to keep the liquid paraffin at the end of the tube

22 The process taking place when liquid paraffin turns to ethene is
A cracking.
B polymerization.
C combustion.
D distillation.
E saponification.

23 At the end of the experiment a liquid was found floating on the water in the trough. This liquid
A is produced by reaction between ethene and water.
B is unreacted liquid paraffin produced by distillation.
C comes from the broken china.
D is produced when ethene comes in contact with the air.
E is produced when ethene reacts with dissolved impurities in the water.

24 Compounds in the same homologous series have the same

- A molecular formula.
- B physical properties.
- C melting and boiling points.
- D empirical formulae.

Classification

Questions 25–30

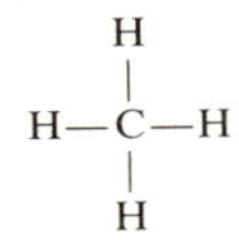

A

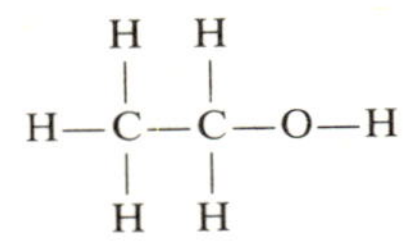

B

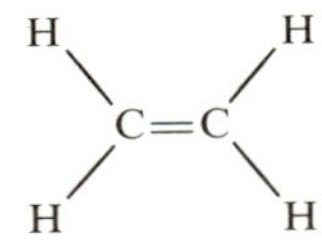

C

H—C≡C—H

D

Which one of the substances A–D

25 is not a hydrocarbon.

26 may be converted into a polymer.

27 has the lowest melting and boiling point.

28 is a member of the alkyne family.

29 reacts with sodium to produce hydrogen.

30 is produced when sodium ethanoate is heated with soda lime.

Multiple completion

Type 1 (See instructions on page 1.)

31 Orange potassium dichromate(VI) crystals are used inside a breathalyzer. When a motorist, who has been drinking to excess, blows through the breathalyzer, the crystals turn green. This is because the
(*i*) ethanol is oxidized. (*ii*) potassium dichromate(VI) is reduced.
(*iii*) chromium(III) sulphate is produced.

32 Cracking of hydrocarbons produces
(*i*) unsaturated products. (*ii*) smaller hydrocarbons. (*iii*) large quantities of energy.

Assertion-reason

(See instructions on pages 1–2.)

33 Ethanol can be distinguished from water by testing the pH.	Ethanol and water are completely miscible.
34 Ethane is a member of the alkane family.	On complete combustion, ethane produces carbon dioxide and water.

Structured questions

35 Figure 33.2 summarizes the relationships between some common compounds. Use this diagram in answering the following questions.

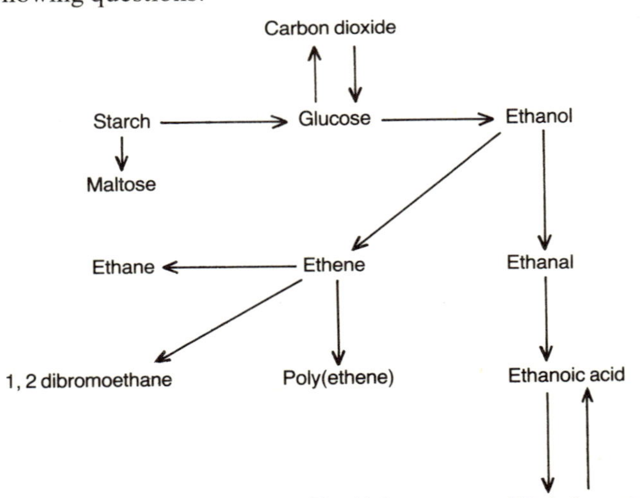

Fig. 33.2

(*a*) From the compounds in Fig. 33.2 select
(*i*) three hydrocarbons, (*ii*) three carbohydrates, (*iii*) an alkene, (*iv*) an alcohol,
(*v*) a compound giving a pH less than 7 when in aqueous solution,
(*vi*) an addition polymer, (*vii*) a condensation polymer, (*viii*) a sweet smelling liquid,
(*ix*) a compound which is produced when wine sours in contact with air,
(*x*) a saturated hydrocarbon.

(*b*) The conversion of starch to poly(ethene) can be brought about in four stages. These four stages involve fermentation, dehydration, polymerization and hydrolysis. Complete the following flow diagram outlining this change. For each step give the type of reaction involved.

starch $\xrightarrow{1}$ [ ] $\xrightarrow{2}$ [ ] $\xrightarrow{3}$ [ ] $\xrightarrow{4}$ poly(ethene)

(*c*) (*i*) What is the usual source of ethene at present?
(*ii*) Describe briefly how ethene is obtained from this source.
(*iii*) Why could starch become the source of poly(ethene) in the future?

34 Carbon dioxide and carbonates

Simple multiple choice

1 When a piece of solid carbon dioxide is left exposed to the air it changes straight into a gas. This process is called
A oxidation. B boiling. C melting. D subliming. E distillation

2 Which one of the following gases does not burn in air?
A carbon monoxide B carbon dioxide C hydrogen D methane

3 Which one of the following oxides will produce a solution of pH less than 7 when shaken with water?
A calcium oxide
B copper (II) oxide
C sodium oxide
D carbon dioxide
E hydrogen oxide

4 Carbon dioxide is prepared by the action of dilute hydrochloric acid on calcium carbonate. The balanced equation for the reaction is
A $CaCO_3(s) + 2HCl(aq) \rightarrow CaCl_2(aq) + H_2(g) + CO_2(g)$.
B $CaCO_3(s) + 2HCl(aq) \rightarrow CaCl_2(aq) + H_2O(l) + CO_2(g)$.
C $CaCO_3(s) + 2HCl(aq) \rightarrow Ca(HCO_3)_2(aq) + HCl(aq) + CO_2(g)$.
D $CaCO_3(s) + 2HCl(aq) \rightarrow CaCl_2(aq) + H_2O(l)$.

5 Refer to Fig. 24.1 on pages 47–8. Which one of the following combinations of apparatus can be used to prepare a dry sample of carbon dioxide from calcium carbonate and dilute hydrochloric acid?
A 2, 4, 6 B 1, 5, 7 C 2, 4, 8 D 3, 4, 8 E 2, 5, 6

6 Which one of the following metal carbonates does **not** evolve carbon dioxide on heating?
A lead(II) carbonate
B copper(II) carbonate
C calcium carbonate
D potassium carbonate
E zinc carbonate

7 Which one of the following compounds reacts with dilute hydrochloric acid to produce a gas which turns limewater milky and leaves a coloured solution?
A chromium(III) carbonate
B lead(II) carbonate
C iron(III) oxide
D ammonium carbonate
E copper(II) oxide

Questions 8–11 refer to the Solvay process.
The process occurs in four steps. The four equations are:
I $CaCO_3(s) \rightarrow CaO(s) + CO_2(g)$
II $NaCl(aq) + NH_3(g) + H_2O(l) + CO_2(g) \rightarrow NaHCO_3(s) + NH_4Cl(aq)$
III $2NaHCO_3(s) \rightarrow Na_2CO_3(s) + CO_2(g) + H_2O(l)$
IV $2NH_4Cl(aq) + Ca(OH)_2(aq) \rightarrow CaCl_2(aq) + 2H_2O(l) + 2NH_3(g)$

8 The Solvay process produces as the main product
A calcium chloride.
B sodium hydrogencarbonate.
C sodium carbonate
D ammonium chloride.
E carbon dioxide.

9 Which compound is precipitated during the process?
A sodium carbonate
B calcium chloride
C ammonium chloride
D sodium hydrogencarbonate
E sodium chloride

10 The process is not used to prepare the corresponding potassium compound because
A potassium hydrogencarbonate is more soluble.
B there is no demand for the corresponding potassium compound.
C potassium chloride is not available as the starting material.
D potassium hydrogencarbonate is not decomposed by heat.

Multiple completion

Type 1 (See instructions on page 1.)

11 $2Mg(s) + CO_2(g) \rightarrow 2MgO(s) + C(s)$
Which of the following statements are correct?
During this reaction
(*i*) carbon dioxide is reduced.
(*ii*) black specks of carbon are formed.
(*iii*) a white solid is produced.

Assertion-reason

(See instructions on pages 1–2.)

12 Carbon dioxide turns limewater milky and then clear again. | Excess carbon dioxide causes calcium hydroxide to reform.

35 Polymers

Simple multiple choice

1 Polymers are built up from small molecules called
A monomers. B alkanes. C allotropes. D units.

2 During the formation of a condensation polymer
A there is a decrease in the volume.
B small molecules such as water molecules are eliminated.
C liquid monomers form a solid polymer.
D the molecules become closely packed together.

3 Polyacrylonitrile is an addition polymer which is prepared from acrylonitrile.

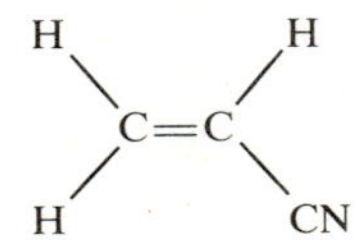

Which one of the following correctly represents the polymer?

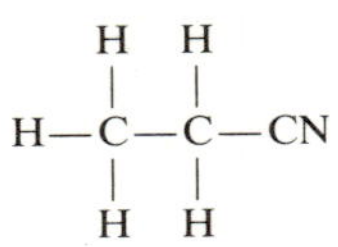
A

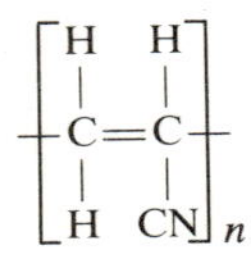
B

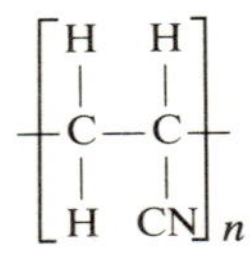
C

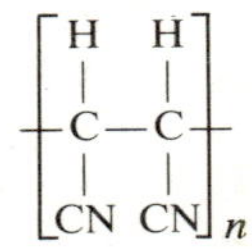
D

4 A plastic which can be softened by heating so that it can be moulded into a particular shape is best called
A a polymer. B thermosetting. C thermoplastic. D thermal.

Classification

Questions 5–8
Below is a list of plastics.
A Polythene
B Perspex
C Nylon
D PVC (polyvinyl chloride)
E Bakelite

Choose the letter which represents the plastic most suitable for making

5 a mountaineer's rope.
6 insulating electrical cables.
7 an ash tray.
8 a flexible food container

Multiple completion

Type 1 (See instructions on page 1.)

9 Poly(phenylethene) is
(*i*) made from polymerizing phenylethene.
(*ii*) a natural polymer. (*iii*) a thermosetting plastic.

Assertion-reason

(See instructions on pages 1–2.)

10 Poly(propene) is an addition polymer. | Propene contains a double bond but poly(propene) contains no double bond.

Structured questions

11

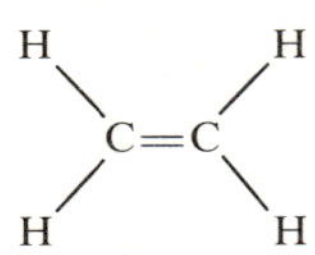
ethene

Ethene can be polymerized under different conditions. If the ethene is heated at high temperatures and pressures in the presence of oxygen, low density poly(ethene) is produced. High density poly(ethene) is produced by bubbling ethene through a heated hydrocarbon solvent containing a dissolved organometallic catalyst.

Low density poly(ethene) is cheaper to produce but has a low melting point and is less strong than high density poly(ethene). High density poly(ethene) is rigid. Both types of poly(ethene)

can easily be moulded using heat and/or vacuum methods.

(*a*) What is the molecular formula of ethene?

(*b*) (*i*) What would be observed if ethene is bubbled through a solution of bromine in hexane?

(*ii*) Write the structural formula of the compound formed when ethene is bubbled through the solution of bromine.

(*c*) Write the structural formula of poly(ethene).

(*d*) Explain why this process is an example of addition polymerization.

(*e*) What is the catalyst in the production of low density poly(ethene)?

(*f*) What is an 'organometallic' catalyst?

(*g*) In which form of poly(ethene) are the molecules more closely packed? Explain your answer.

(*h*) Which of the forms of poly(ethene) is most suitable for each of the following uses? Give a reason for your answer in each case.

(*i*) washing up bowl (*ii*) plastic bag for sandwiches (*iii*) milk crates

(*i*) The polymer poly(propene) is often used instead of high density poly(ethene). Write the structural formula of the monomer of poly(propene).

36 Fuels

Simple multiple choice

1 Which one of the following is not a solid fuel?
A peat B wood C anthracite D naphtha E coal

2 Lead tetraethyl is added to petrol to
A increase flammability.
B improve the burning of the petrol.
C increase the boiling point.
D make the petrol more volatile.

3 The chief constituent of coal gas is
A methane. B carbon dioxide. C hydrogen. D ethene.

4 The chief constituent of natural gas is
A methane.
B helium.
C hydrogen sulphide.
D hydrogen.
E carbon monoxide.

5 The gas present in producer gas which burns is
A nitrogen.
B carbon monoxide.
C methane.
D hydrogen.
E carbon dioxide.

6 The burning of paraffin in a plentiful supply of air produces
A carbon monoxide and hydrogen.
B carbon monoxide and water.
C carbon and water.
D carbon dioxide and hydrogen.
E carbon dioxide and water.

Questions 7 and 8 refer to Fig. 36.1. This apparatus was used to investigate the products of burning different gaseous fuels.

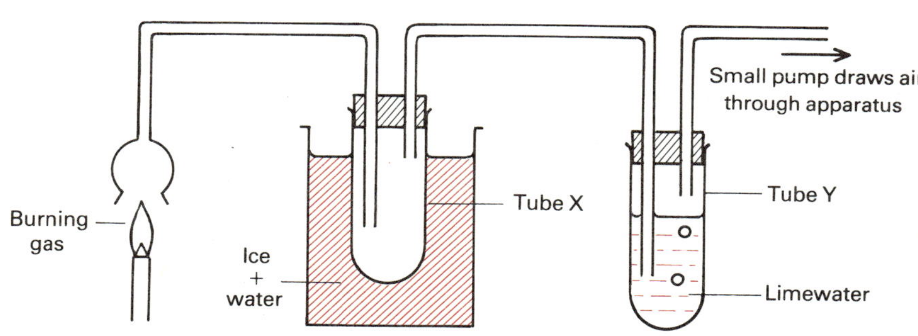

Fig. 36.1

7 When the gas I was burnt in the apparatus, with the pump running, a colourless liquid was collected in X but no change was seen in tube Y within 5 minutes. The liquid collected was found to boil at 100°C. Gas I was, therefore,
A carbon monoxide.
B hydrogen sulphide.
C methane
D hydrogen.
E carbon dioxide.

8 The experiment was repeated with gas II burnt in the apparatus. The same liquid was collected in X and the limewater in Y turned milky. Gas II could be

A carbon monoxide.
B hydrogen.
C methane.
D carbon dioxide
E nitrogen.

Multiple completion

Type 1 (See instructions on page 1.)

9 Products of the destructive distillation of coal include
(*i*) smokeless fuel. (*ii*) coal gas. (*iii*) ammonia.

Assertion-reason

(See instructions on pages 1–2.)

10	Coal, oil and gas are called fossil fuels.	As fossil fuels are used up they can readily be replaced in the earth.

37 Energy in chemistry

Simple multiple choice

1 An exothermic reaction is a reaction in which
A electrical energy is produced.
B the products contain less energy than the reactants.
C energy is necessary to start the reaction.
D the reaction vessel feels cold to the touch as the reaction proceeds.

2 Which one of the following processes is endothermic?
A the burning of petrol
B dissolving ammonium nitrate in water
C the reaction of an acid and an alkali
D the reaction of alkali metals with water
E adding concentrated sulphuric acid to water

3 A fuel cell is a device for
A burning a fuel efficiently.
B converting chemical energy into electrical energy efficiently.
C converting mechanical energy into electrical energy efficiently.
D turning electricity into a convenient liquid fuel.

4 From the following information, select the substance which gives the most energy per gram on combustion.

Substance	*Mass of 1 mole (g)*	*Energy per mole (kJ per mole)*
A	120	5500
B	16	880
C	2	285
D	240	2730
E	12	395

5 When 100 g of water was heated for 4 minutes using a Bunsen burner the temperature rose by 50°C. Assuming the Bunsen burner is supplying energy at a constant rate, how much energy was supplied per minute?
(4.2 J raise the temperature of 1 g of water by 1°C.)
A 8.4 kJ B 1.19 kJ C 5.25 kJ D 21 kJ E 84 kJ

6 Figure 37.1 shows the energy changes occurring during an exothermic reaction.

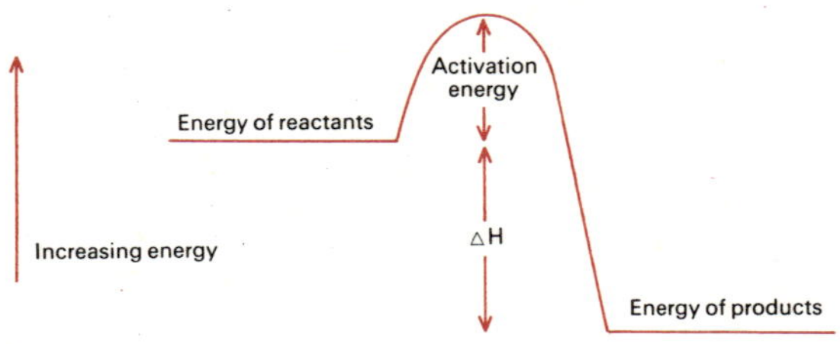

Fig. 37.1

Which one of the following statements about an exothermic reaction is **incorrect?**

A Energy is given out during the reaction.
B The energy of the products is less than that of the reactants.
C ΔH represents the energy evolved during the reaction.
D The products possess greater energy than the reacting substances.
E The activation energy is necessary to start the reaction.

7 When 1 g of carbon is burnt in excess oxygen 33 kJ of energy is released. Calculate the heat of combustion of carbon. ($A_r(C) = 12$)

A 33 kJ per mole. C 33×12 kJ per mole. E $\frac{33 \times 12}{1000}$ kJ per mole.

B $\frac{33}{12}$ kJ per mole D $\frac{12}{33}$ kJ per mole

Multiple completion

Type 2 (See instructions on page 1.)

8 Which of the following processes are exothermic?

(*i*) $CaCl_2(s) \rightarrow Ca^{2+}(l) + 2Cl^-(l)$
(*ii*) $C(s) + H_2O(g) \rightarrow CO(g) + H_2(g)$
(*iii*) $H_2(g) \rightarrow 2H(g)$
(*iv*) $2H_2(g) + O_2(g) \rightarrow 2H_2O(l)$

9 The heat of neutralization of a strong acid and a strong alkali is −56 kJ per mole? In which of the following reactions would the heat evolved **not** be 56 kJ, assuming one mole of the acid reacts in each case?

(*i*) $H^+(aq) + OH^-(aq) \rightarrow H_2O(l)$
(*ii*) $CH_3COOH(aq) + NaOH(aq) \rightarrow CH_3COONa(aq) + H_2O(l)$
(*iii*) $H_2SO_4(aq) + NaOH(aq) \rightarrow NaHSO_4(aq) + H_2O(l)$
(*iv*) $H_2SO_4(aq) + 2NaOH(aq) \rightarrow Na_2SO_4(aq) + 2H_2O(l)$

Assertion-reason

(See instructions on pages 1–2.)

10 The products of an exothermic reaction contain more energy than the reactants at the same temperature and pressure. — During an exothermic reaction energy is evolved.

Structured questions

11 In a neutralization experiment 500 cm^3 of 2 M hydrochloric acid (2 moles per dm^3) and 500 cm^3 of 2 M sodium hydroxide solution were mixed in a poly(phenylethene) container with a lid. The temperature of both solutions before mixing was 20°C and after mixing the temperature rose to 33.0°C.

(*a*) Write a symbol and ionic equation for the reaction taking place.
(*b*) What is the advantage of using poly(phenylethene) rather than metal for the reaction container?
(*c*) What is the temperature rise on mixing the solutions?
(*d*) What is the total volume of the solution on mixing?
(*e*) Assuming 1 cm^3 of this solution weighs 1 g, what is the mass of the final solution?
(*f*) Calculate the heat evolved on mixing these solutions.
(The heat required to raise the temperature of 1 kg of solution by 1°C is 4.2 kJ.)
(*g*) Why is the experimental result slightly lower than the value given in a textbook?

38 Social and economic considerations

Simple multiple choice

1 Many plastics cause a pollution problem because they are long-lasting. The chief reason for their pollution threat is that they are

A very strong and tough. B resistant to bacteria and chemical action.
C composed of carbon. D very widely used.

2 Sulphur dioxide is a major cause of air pollution. Which one of the following properties of sulphur dioxide could be used to reduce the amount of sulphur dioxide leaving a factory chimney? Sulphur dioxide

A is soluble in water.
B is denser than air.
C turns limewater milky.
D has a pungent odour.
E is liquefied under pressure.

3 Which one of the following elements present in coal produces a pollution problem when coal is burnt?

A carbon B sulphur C hydrogen D silicon E nitrogen

4 Which one of the following materials **cannot** easily be recycled?
A copper
B glass
C aluminium
D thermosetting plastics
E thermoplastics

5 Which one of the following gases in the air seriously pollutes the atmosphere?
A carbon dioxide
B nitrogen
C argon
D water vapour
E sulphur dioxide

6 Which one of the following gases in the air of an industrial city is likely to darken white lead-based paints?
A carbon monoxide
B hydrogen sulphide
C sulphur dioxide
D nitrogen dioxide
E carbon dioxide

7 Traffic police on point-duty in Tokyo, Japan, use breathing apparatus because of the harmful effects of
A carbon dioxide.
B sulphur dioxide.
C water vapour.
D hydrogen sulphide.
E carbon monoxide.

Multiple completion

Type 1 (See instructions on page 1.)

8 The pollution of water with waste ammonia may
(*i*) remove oxygen dissolved in the water.
(*ii*) cause green algae to grow on the surface.
(*iii*) produce a pH less than 7.

9 When setting up a factory to produce sodium by electrolysis it is important to
(*i*) site it well away from populated areas because of the danger of the process.
(*ii*) have a supply of sodium chloride from sea or salt deposits.
(*iii*) have a firm wishing to buy waste chlorine.

Assertion-reason

(See instructions on pages 1–2.)

10 Nitrogen is a pollutant from a factory using the Haber process. | The Haber process is a process for converting nitrogen and hydrogen into ammonia.

39 Radioactivity

Simple multiple choice

1 Which one of the following consists of a stream of negatively charged particles?
A α-particles B β-particles C X-rays D γ-rays

2 A radioactive source emits α, β and γ radiation. Which of the following types of radiation will be detected if a piece of paper is placed between the source and the detector?
A α only B α and β C γ only D β and γ

3 At the start of an experiment the radioactive sample gave a corrected count of 120 counts per minute. After 3 hours the count had dropped to 15 counts per minute (after correction). The half life of the sample was
A 30 minutes.
B 45 minutes.
C 60 minutes.
D 75 minutes.
E 90 minutes.

4 In which one of the following ways does radioactive decay differ from chemical reaction? In radioactive decay
A no new substances are produced.
B energy is evolved.
C the rate of reaction depends upon the amount present.
D the rate of reaction is not changed by a change of temperature.
E the products are charged particles.

For *questions 5–8* refer to the following list of symbols and atomic numbers for some of the heavier elements.

Atomic numbers	82	83	84	85	86	87	88	89	90	91
Symbol	Pb	Bi	Po	At	Rn	Fr	Ra	Ac	Th	Pa

Below is part of a radioactive series.

$^{233}_{91}\text{Pa} \xrightarrow{\beta} \text{X} \xrightarrow{\alpha} \text{Y} \xrightarrow{\alpha} \text{Z} \xrightarrow{\beta} {}^{225}_{89}\text{Ac}$

5 Which of the following elements is a member of the alkali metal family?
A At   B Rn   C Fr   D Ra   E Pa

6 Which one of the following species corresponds to X?
A $^{233}_{90}\text{Th}$ B $^{233}_{92}\text{U}$ C $^{229}_{88}\text{Ra}$ D $^{229}_{89}\text{Ac}$ E $^{231}_{89}\text{Ac}$

7 Which one of the following species corresponds to Y?
A $^{229}_{90}\text{Th}$ B $^{230}_{91}\text{Pa}$ C $^{228}_{88}\text{Ra}$ D $^{229}_{89}\text{Ac}$ E $^{229}_{91}\text{Pa}$

8 Which one of the following species corresponds to Z?
A $^{227}_{88}\text{Ra}$ B $^{225}_{89}\text{Ac}$ C $^{226}_{89}\text{Ac}$ D $^{225}_{88}\text{Ra}$ E $^{228}_{89}\text{Ac}$

Multiple completion

Type 1 (See instructions on page 1.)

9 Radioactive decay involves
(*i*) the formation of a new species.   (*ii*) the emission of charged particles.
(*iii*) the giving out of energy.   (*iv*) a decrease in the volume of the substance.

10 Which of the following statements about the half life of a radioactive isotope is (are) correct?
(*i*) All isotopes of the same element have the same half life.
(*ii*) All of the isotope decays in two half lives.
(*iii*) the half life of an isotope remains the same even if other substances are mixed with it.

Structured questions

11 Bismuth-212 is a radioactive isotope with a half life of one hour. In an experiment the background count was found to be 20 counts per minute. At the start of the experiment the uncorrected count for the sample of bismuth-212 was 532 counts per minute.
(*a*) Explain briefly how the average background count could be found.
(*b*) What is the corrected count for the first minute?
(*c*) Plot a graph of the corrected count after 1 hour, 2 hours, 3 hours, 4 hours and 5 hours against time. Draw a curve through these points.
(*d*) Bismuth-212 decays by β-decay to produce X which in turn decays to produce Y by α-decay. Using the table for questions 5–8 on p. 66, identify the mass number, atomic number and name of X and Y respectively.

40 Quantitative volumetric chemistry

Simple multiple choice

1 Which one of the following pieces of apparatus would be most suitable for adding exactly 24.5 cm^3 of dilute acid to a flask?
A pipette   D measuring cylinder
B gas syringe   E teat pipette
C burette

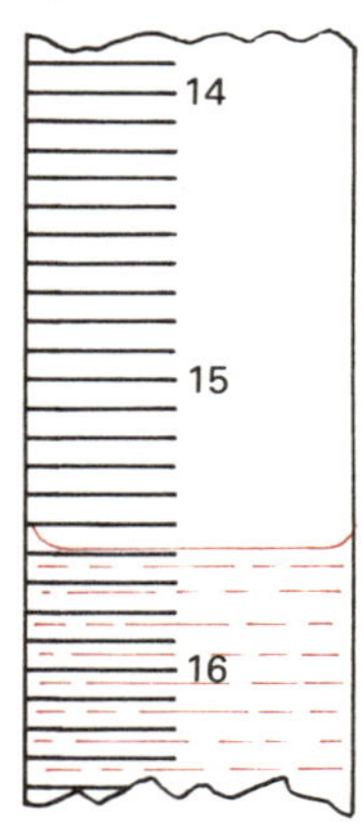

Fig. 40.1

2 A pipette cannot be used for accurately measuring out the volumes of hot liquids because the
A pipette would crack.   B pipette would be too hot to hold.
C liquids would be expanded.   D temperature would fall during the operation.

3 The accurate reading on the burette in Fig. 40.1 is
A 16.25 cm^3.   B 15.55 cm^3.   C 15.70 cm^3 D 15.75 cm^3.   E 15.80 cm^3.

4 Which one of the following solutions contains the greatest number of moles?
A 50 cm^3 of 0.5 M hydrochloric acid   D 100 cm^3 of M hydrochloric acid.
B 100 cm^3 of 0.5 M sodium hydroxide   E 50 cm^3 of 5 M nitric acid
C 50 cm^3 of M sulphuric acid

5 $H_2SO_4(aq) + 2NaOH(aq) \rightarrow Na_2SO_4(aq) + H_2O(l)$
The volume of M sulphuric acid required to react with 50 cm^3 of M sodium hydroxide solution according to the above equation is
A 12.5 cm^3.   B 25 cm^3.   C 50 cm^3.   D 100 cm^3.
E none of these answers.

6 Which one of the following procedures is **not** correct when titrating an acid into an alkali?
A Wash out the burette with acid and fill the burette with acid.
B Wash out the pipette with alkali.
C Rinse out the flask with alkali and add exactly 25 cm^3 of alkali from a pipette.
D Add a couple of drops of indicator to the contents of the flask.
E Read the volume of acid in the burette before and after titrating.

Classification

Questions 7–9

These questions relate to the table below.

Indicator	*Colour in dilute hydrochloric acid*	*Colour in pure water*	*Colour in sodium hydroxide solution*
Screened methyl orange	Red	Green	Green
Phenolphthalein	Colourless	Colourless	Magenta
Bromothymol blue	Yellow	Green	Blue

Table 40.1

Using the table above classify the substances in questions 7–9 according to the following:

A strongly acidic
B weakly acidic
C neutral
D weakly alkaline
E strongly alkaline

7 A solution which is blue/green with bromothymol blue and colourless with phenolphthalein.

8 A solution which is green with screened methyl orange and yellow with bromothymol blue.

9 A solution which is green with screened methyl orange and colourless with phenolphthalein.

Assertion-reason

(See instructions on pages 1–2.)

10 During the titration of an acid with an alkali the conductivity of the solution rises. | Both acid and alkali are composed of ions.

Structured questions

11 Washing soda is hydrated sodium carbonate $Na_2CO_3.xH_2O$. An experiment was carried out to find the value of *x*.

28.6 g of fresh hydrated sodium carbonate crystals were dissolved in distilled water and the solution made up to a volume of 1000 cm^3. 25 cm^3 of this solution was transferred to a conical flask and a few drops of methyl orange indicator added.

The solution in the conical flask was titrated with 0.25 M hydrochloric acid (0.25 mole per dm^3) and 20.0 cm^3 of acid were required to reach the end point.

(*a*) Name the most suitable piece of apparatus for measuring out exactly 25 cm^3 of sodium carbonate solution and transferring it to the conical flask.
(*b*) How would you know that the end point of the titration had been reached?
(*c*) Write an equation for the reaction of sodium carbonate with dilute hydrochloric acid using methyl orange as indicator.
(*d*) How many moles of hydrochloric acid react with 25 cm^3 of sodium carbonate solution?
(*e*) How many moles of hydrochloric acid would react with 1000 cm^3 of sodium carbonate solution?
(*f*) Using the equation, how many moles of anhydrous sodium carbonate are present in 1000 cm^3 of sodium carbonate solution?
(*g*) What mass of anhydrous sodium carbonate is present in 1000 cm^3 of sodium carbonate solution?
($A_r(Na)=23$, $A_r(C)=12$, $A_r(O)=16$)
(*h*) Complete the following:
____ g of anhydrous sodium carbonate combines with ____ g of water to produce 28.6 g of sodium carbonate crystals.
(*i*) How many grams of water would combine with one mole of sodium carbonate?
(*j*) What is the value of *x*? ($A_r(H)=1$, $A_r(O)=16$)

41 Qualitative chemistry

Simple multiple choice

1 Calcium can be detected using a flame test because calcium compounds colour the flame
A red. B green. C orange. D lilac. E blue.

2 Which one of the following would form a white precipitate with an aqueous solution of silver nitrate acidified with dilute nitric acid?
A sulphates B carbonates C nitrates D chlorides E iodides

3 Which one of the following metal ions in solution will **not** give a precipitate with sodium hydroxide solution?
A $Cu^{2+}(aq)$ B $Fe^{2+}(aq)$ C $Na^{+}(aq)$ D $Fe^{3+}(aq)$ E $Pb^{2+}(aq)$

4 A metal hydroxide which is insoluble in water but soluble in a solution of sodium hydroxide or dilute sulphuric acid could be a hydroxide of
A aluminium. B sodium. C calcium. D magnesium. E copper.

5 When an aqueous solution of sodium hydroxide is added to a freshly prepared solution of iron(II) sulphate solution
A a colourless solution is formed.
B a colourless precipitate is formed.
C a coloured solution is formed.
D a green precipitate is formed.
E a red-brown precipitate is formed.

6 Which one of the following evolves brown fumes when dilute hydrochloric acid is added?
A sodium bromide
B sodium nitrate
C sodium hydroxide
D sodium nitrite
E sodium sulphite

Classification

Questions 7–10
A white precipitate
B colourless solution
C colourless gas and colourless solution
D colourless gas and coloured solution
E coloured precipitate

For questions **7–10** select the letter A–E above which best describes what would be observed in each reaction.

7 sodium hydroxide solution added to copper(II) sulphate solution

8 sodium chloride solution added to an aqueous solution of silver nitrate acidified with dilute nitric acid

9 dilute sulphuric acid added to chromium(III) carbonate

10 ammonia solution added to iron(III) chloride solution

Multiple completion

Type 1 (See instructions on page 1.)

11 Which of the following would evolve a colourless gas on addition of dilute hydrochloric acid and without heating?
(*i*) a carbonate (*ii*) a hydrogencarbonate (*iii*) a nitrate

Structured questions

12 Calamine is a mineral. When crushed, calamine is added to dilute hydrochloric acid. Effervescence is observed and a colourless, odourless gas A is evolved. The gas extinguishes a lighted splint and turns limewater milky. A colourless solution B remains.

On heating calamine turns to a yellow powder and the gas A is again evolved. On cooling the yellow powder turns to a white powder C.

(*a*) Identify
(*i*) the gas A, (*ii*) the colourless solution B, (*iii*) the white powder C.
(*b*) What is the chief chemical constituent of calamine?
(*c*) Calamine in suspension in water can be used on insect bites and stings. Explain why it is so used.
(*d*) Suggest a method of obtaining a metal from calamine.

Answers

1 Elements, mixtures and compounds

1 C **2** D **3** C **4** A **5** D **6** D **7** B **8** E **9** A **10** D **11** E **12** B **13** C
14 D **15** A **16** E **17** B **18** A **19** A **20** B **21** D **22** B

2 Separation techniques in chemistry

1 B **2** A **3** C **4** D **5** B **6** B **7** C **8** A **9** B **10** E **11** D **12** C **13** D
14 C **15** D **16** A **17** C **18** B **19** C **20** A **21** A **22** E **23** A **24** B

25 (*a*) To assist in the crushing of plant cells. (*b*) Immerse the boiling tube in a beaker of hot water. No flames. (*c*) Flask with condenser arranged vertically above. (*d*) Filtration (or decanting or centrifuging). (*e*) (*i*) Two circles one inside the other: green ring on the inside; yellow ring on the outside. (*ii*) Cut out a piece of filter paper containing part of the green ring and dissolve the dye in propanone. Similar for yellow dye. (*f*) In spring more chlorophyll is present than xanthrophyll. The situation is reversed in autumn. (*g*) See Fig. 2.1.

3 Symbols, formulae and equations

1 A **2** B **3** C **4** E **5** A **6** C **7** B **8** D **9** E **10** C **11** D **12** C **13** E
14 A **15** D **16** B **17** C **18** A **19** C **20** B **21** A

4 Structure of the atom

1 D **2** B **3** C **4** B **5** B **6** C **7** A **8** E **9** E **10** B **11** C **12** B **13** D
14 D **15** A **16** E **17** D **18** B **19** A **20** C **21** C **22** B **23** B **24** E
25 B **26** C **27** D **28** A **29** D

5 Bonding and structure

1 A **2** C **3** B **4** A **5** D **6** A **7** E **8** A **9** B **10** C **11** E **12** C **13** E
14 B **15** A **16** D **17** D **18** A **19** D **20** E **21** C **22** D **23** C **24** A

25 (*a*) (*i*) RP (it is usual to put the metal first). (*ii*) QP_3 (*iii*) R_3Q
(*b*) (*i*) Covalent (*ii*) Ionic (*iii*) Covalent
(*c*) (*i*) P, Q (*ii*) R

6 States of matter

1 D **2** D **3** D **4** C **5** C **6** D **7** A **8** E **9** D **10** B **11** A **12** C **13** B
14 B **15** D **16** D **17** A **18** C **19** E **20** B **21** C **22** B **23** A **24** C
25 D **26** E **27** E **28** B **29** A **30** B **31** C **32** C **33** A **34** A **35** B **36** C

37 (*a*) Ammonium chloride (*b*) Ammonia particles move faster than hydrogen chloride particles. (*c*) The movement of particles is not in one direction but random. The tube is filled with air. Collisions between particles in air and ammonia or hydrogen chloride particles. (*d*) Diffusion (*e*) Sulphur (*f*) The ring of sulphur is formed closer to the pad of cotton wool soaked in sulphur dioxide solution. Hydrogen sulphide particles ($M_r = 34$) are smaller and lighter than sulphur dioxide particles ($M_r = 64$). Hydrogen sulphide particles, therefore, move faster.

7 Some laws relating to gases

1 B **2** B **3** A **4** D **5** D **6** B **7** A **8** A **9** E **10** D **11** B **12** C **13** A
14 E

8 Air and combustion

1 B **2** B **3** A **4** D **5** D **6** B **7** A **8** D **9** D **10** C **11** E **12** D **13** D
14 E **15** B **16** D **17** C **18** A **19** B **20** B **21** C **22** E **23** D **24** A **25** D

9 Oxygen and oxides

1 D **2** E **3** C **4** B **5** B **6** C **7** A **8** B **9** C **10** A **11** D **12** B **13** E
14 A **15** C **16** E **17** D **18** B **19** E **20** A

10 Hydrogen

1 C **2** B **3** A **4** B **5** C **6** E **7** C **8** D **9** B **10** A

11 (*a*) Atoms of the same element containing different numbers of neutrons. They have the same atomic numbers but different mass numbers. (*b*) Hydrogen: mass number = 1 Deuterium: mass number = 2 Tritium: mass number = 3 (*c*) Hydrogen is a mixture of three isotopes. The relative atomic mass is the average of the mass numbers of the three isotopes related to the percentage of each isotope present.

(*d*) H^+ D^- D_2

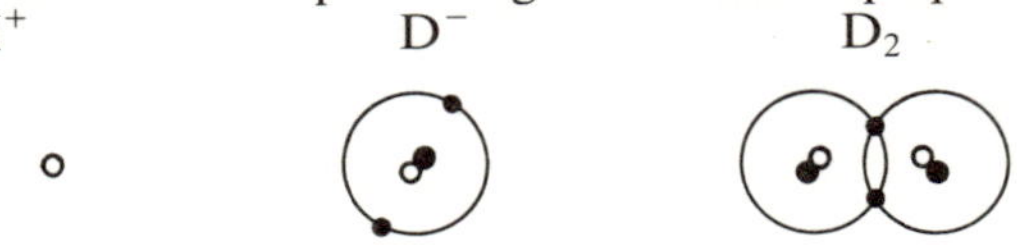

(*e*) Deuterium and tritium are present in such small amounts. (*f*) Both methods require a great deal of time and energy. (*g*) Boiling point. (*h*) Add a piece of calcium to heavy water and collect the deuterium produced over water or by upward delivery.

$$Ca(s) + 2D_2O(l) \rightarrow Ca(OD)_2(s) + D_2(g)$$

Note: Do not use potassium or sodium as they are too reactive.

(*i*) This is not as difficult as it seems. It is just the test for hydrogen. Put a lighted splint into the gas: deuterium burns with a squeaky pop and the splint is extinguished.

(*j*) (*i*) $CuO(s) + D_2(g) \rightarrow Cu(s) + D_2O(g)$
(*ii*) $N_2(g) + 3D_2(g) \rightleftharpoons 2ND_3(g)$ (*iii*) $D_2(g) + Cl_2(g) \rightarrow 2DCl(g)$

11 Water

1 E **2** A **3** B **4** A **5** C **6** C **7** E **8** B **9** C **10** A **11** B **12** D **13** A
14 E **15** B **16** E **17** D **18** C **19** A **20** D **21** E **22** C **23** B **24** B
25 B **26** A **27** B **28** A **29** C **30** B **31** A **32** A

33 (*a*) The solubility of X in water increases with rise in temperature.
(*b*) 36 g per 100 g of water (*c*) (*i*) Y; (*ii*) X (*d*) 38.5°C. (*e*) Solubility 48 g per 100 g of water. Temperature = 50°C. (*f*) (*i*) 12 g of X remained undissolved. (Only 68 g of X can dissolve in 100 g of water at 60°C.) No Y remains undissolved. (40 g of Y can be dissolved in 100 g of water at 60°C.)
(*ii*) 19 g of X can be dissolved in 100 g of water at 15°C. Since 12 g of X had already been removed at 60°C (68-19)g = 49 g of X crystallize out of solution. 29 g of Y can be dissolved in 100 g of water at 15°C. 11 g of Y crystallize out of solution.

34 (*a*) Teat pipette (*b*) Calcium and magnesium ions. Any salt containing calcium or magnesium ions forms only a little lather with soap (*c*) (*i*) Measuring cylinder (*ii*) Burette
(*d*) Temporary hardness – completely removed by boiling. (*e*) Calcium hydrogencarbonate
(*f*) $CaCO_3(s) + H_2O(l) + CO_2(g) \rightleftharpoons Ca(HCO_3)_2(aq)$
(*g*) The volume of soapless detergent required to produce a lather is the same in hard and soft water. All volumes in the table would be the same.

12 The mole and chemical calculations

1 C **2** C **3** A **4** B **5** E **6** D **7** D **8** B **9** B **10** D **11** B **12** D **13** A
14 C **15** D **16** E **17** C **18** B **19** D **20** A **21** B **22** C **23** B **24** A
25 D **26** C **27** D **28** B

29 (*a*) Black to pinkish-brown
(*b*) Mixtures of hydrogen and air are explosive. Hydrogen pushes the air out of the apparatus and the gas can be burnt safely.
(*c*) If air comes into contact with hot copper, the copper is oxidized to copper(II) oxide.
(*d*) 0.19 g, 0.20 g
(*e*) Graph. The line does not pass through all points due to experimental error.
(*f*) 0.16 g
(*g*) $0.16 \times 100 = 16$ g (*h*) $\frac{128}{64} = 2$ moles (*i*) $\frac{16}{16} = 1$ mole
(*j*) Cu_2O. This is copper(I) oxide. It is not the usual oxide of copper and this question demonstrates that it is unwise to assume that you know what answer to expect.

13 Laws of chemical combination

1 B **2** C **3** B **4** A **5** C **6** E **7** A **8** B **9** B **10** A

14 Metals(I)

1 C **2** A **3** D **4** C **5** A **6** E **7** D **8** D **9** B **10** D **11** C **12** E **13** A
14 C **15** D **16** A **17** B **18** E

19 (*a*) An alloy is a mixture of metals (or metals and carbon). (*b*) Brass, duralumin, steel.
(*c*) Nitrogen dioxide (*d*) This part of the experiment should be carried out in a fume cupboard. The fumes are very harmful to the lungs. (*e*) A mixture of the nitrates of the constituent metals plus excess nitric acid. (*f*) German silver contains zinc, copper and nickel (no silver). (*g*) Paper chromatography (*h*) Cheap jewellery – looks like silver but cheaper.

15 Metals(II)

1 E **2** C **3** B **4** B **5** A **6** D **7** D **8** A **9** E **10** E **11** D **12** C **13** D
14 E **15** A **16** D **17** C **18** C **19** D **20** A **21** A **22** B **23** E **24** C
25 A

26 (*a*) Aluminium is only present in clay in very small amounts. It is not economic, at present, to consider extracting aluminium from clay. (*b*) The extraction of aluminium is costly because of the large amounts of electricity required. The extraction of iron is a reduction process with coke and does not use electricity. (*c*) Amphoteric (*d*) Filtration (*e*) To assist with the tapping of aluminium from the furnace. A small amount of aluminium will collect close to the tap. (*f*) The high temperature required to melt aluminium oxide would cause considerable extra fuel expense. Also, aluminium oxide has considerable covalent tendencies and molten aluminium oxide would not conduct electricity well.
(*g*) Cathode $Al^{3+} + 3e^- \rightarrow Al$
Anode $2O^{2-} \rightarrow O_2 + 4e^-$
(*h*) Carbon anodes burn in oxygen produced at the temperature of the furnace.
$C(s) + O_2(g) \rightarrow CO_2(g)$
(*i*) Cheaper electricity is available from hydroelectric power.

(*j*) Sample answers only. There are alternatives. (*i*) Good conductor of electricity (*ii*) Low density for metal (*iii*) Malleable – can be made into thin sheets (*iv*) Good conductor of heat (*v*) Strength

(*k*) (*i*) Low density reduces the mass of the ship. This increases speed, reduces fuel consumption, enables extra equipment to be incorporated and allows hull shapes to be constructed which would otherwise be impossible. (*ii*) Aluminium is a reactive metal (high in the reactivity series). It is not rapidly corroded or apparently reactive because of its oxide layer. Once aluminium catches alight it burns rapidly.

16 Oxidation and reduction

1 C **2** D **3** E **4** A **5** C **6** B **7** D **8** E **9** D **10** B

11 (*a*) (*i*) Dilute hydrochloric acid (*ii*) Neither oxidation nor reduction
(*iii*) $Na_2CO_3(s) + 2HCl(aq) \rightarrow 2NaCl(aq) + H_2O(l) + CO_2(g)$

(*b*) (*i*) Chlorine (*ii*) Oxidation
(*iii*) $2FeCl_2(aq) + Cl_2(g) \rightarrow 2FeCl_3(aq)$

(*c*) (*i*) Copper(II) oxide (*ii*) Oxidation
(*iii*) $2NH_3(g) + 3CuO(s) \rightarrow 3Cu(s) + 3H_2O(g) + N_2(g)$

(*d*) (*i*) Sodium hydroxide (or calcium hydroxide) (*ii*) Neither oxidation nor reduction
(*iii*) $NH_4Cl(s) + NaOH(s) \rightarrow NaCl(s) + H_2O(g) + NH_3(g)$

(*e*) (*i*) Chlorine (*ii*) Oxidation
(*iii*) $2KI(aq) + Cl_2(g) \rightarrow 2KCl(aq) + I_2(aq)$

17 The effect of electricity on chemicals

1 B **2** A **3** A **4** D **5** D **6** C **7** B **8** A **9** B **10** E **11** D **12** A **13** B
14 B **15** D **16** D **17** A **18** B **19** C **20** A **21** D **22** E **23** D **24** B
25 B **26** C **27** A

28 (*a*) (*i*) Electroplating (*ii*) The copper disc is the cathode and a nickel rod is the anode. These are connected in series with a battery, a rheostat (to adjust the current flowing) and an ammeter (to measure the current flowing). The two electrodes dip into nickel(II) sulphate solution.
(*iii*) Adjusting the rheostat.
Moving the electrodes further apart.

(*b*) (*i*) Copper (not rust which is frequently given on examination scripts).
(*ii*) Nickel is above copper but below iron in the reactivity series. Nickel will replace copper from copper(II) sulphate but not iron from iron(II) sulphate.

(*c*) (*i*) To prevent corrosion. Nickel is less reactive than iron.
(*ii*) Chromium (not chrome). Chromium can be polished to give a much greater shine.

18 Chemical families and the Periodic Table

1 E **2** B **3** C **4** A **5** B **6** C **7** C **8** C **9** C **10** D **11** A **12** B **13** C
14 C **15** E **16** D **17** A **18** E **19** A **20** E **21** A **22** E **23** B **24** D
25 A **26** D **27** E **28** C **29** C **30** A

31 (*a*) (*i*) $B_2O_3(s) + 3Mg(s) \rightarrow 3MgO(s) + 2B(s)$
(*ii*) Remove excess magnesium and magnesium oxide by adding dilute hydrochloric acid.

$Mg(s) + 2HCl(aq) \rightarrow MgCl_2(aq) + H_2(g)$
$MgO(s) + 2HCl(aq) \rightarrow MgCl_2(aq) + H_2O(l)$

Boron is removed by filtering and then washed with distilled water and dried.

(*b*) (*i*) Catalyst
(*ii*) By passing an electric current through it
(*iii*) Reducing agent
(*iv*) $2BF_3(g) + 3H_2(g) \rightarrow 2B(s) + 6HF(g)$

(*c*) (*i*) $2B(s) + 3Cl_2(g) \rightarrow 2BCl_3(g)$
(*ii*) Covalent

(*d*) B_5H_9. Carbon forms a wide range of hydrocarbons.

(*e*) $Mg_3B_2(s) + 6HCl(aq) \rightarrow 3MgCl_2(aq) + B_2H_6(g)$

(*f*) 5 electrons because the atomic number is 5. The electron arrangement is 2,3.

(*g*) Group III (3 electrons in the outer shell). Period 2.

19 Acids, bases and salts

1 D **2** A **3** C **4** C **5** B **6** C **7** D **8** B **9** D **10** D **11** D **12** D **13** D
14 B **15** A **16** C **17** C **18** A **19** D **20** B **21** A **22** D **23** B **24** D
25 D **26** A **27** C **28** E **29** A **30** A

31 (*a*) A sodium nitrate
B water
C sulphuric acid
D hydrogen
E barium chloride (or barium nitrate)
F sodium chloride (or sodium nitrate)
G iron
H chlorine
I lead(II) carbonate
J nitric acid

(*b*) $CuO(s) + H_2SO_4(aq) \rightarrow CuSO_4(aq) + H_2O(l)$
(*c*) (*i*) thermometer; (*ii*) spatula; (*iii*) evaporating basin
(*d*) All of the acid is used up.

Note: A frequent mistake on examination scripts is the assumption that copper(II) oxide remains unchanged in the solution because the solution is saturated with copper(II) oxide. There is confusion between reaction and dissolving. (*e*) Filtration (*f*) The solution changes from colourless to blue. (*g*) On heating to dryness, the water of crystallization is lost and anhydrous copper(II) sulphate is formed.

$CuSO_4.5H_2O(s) \rightleftharpoons CuSO_4(s) + 5H_2O(g)$

20 Rates of reaction

1 B 2 C 3 D 4 A 5 B 6 C 7 A 8 B 9 C 10 D 11 B 12 C 13 A
14 C 15 D 16 C 17 C 18 D 19 A 20 B

21 (*a*) Calcium carbonate (*b*) Carbon dioxide escapes from the solution. (*c*) Cotton wool prevents the escape of acid spray. (It is not there to prevent the escape of carbon dioxide.)
(*d*) Graph – similar in shape to Fig. 20.2 (*e*) At the start of the reaction. (*f*) 10 minutes.
(*g*) Since some marble remains after the reaction is complete, the marble must be in excess, i.e. all the acid is used up. (*h*) Because the reaction is so fast, it is complete almost before the apparatus can be assembled. Figure 20.1 in the *Revise Chemistry* O-level guide shows a possible method of carrying out the experiment by keeping the reactants apart during the assembly of the apparatus. (*i*) Similar in shape to the graph plotted but: the graph is steeper than the graph plotted; the graph reaches a maximum at 20 cm^3 in a time less than 10 minutes.

21 Reversible reactions and equilibrium

1 D 2 A 3 B 4 B 5 A 6 A 7 E 8 A 9 C 10 B 11 D 12 A 13 C
14 E 15 B 16 A

17 (*a*) Pass steam over heated iron in a combustion tube. Collect the hydrogen over water. (*Revise Chemistry* O-level guide, Fig. 21.1.) (*b*) Pass hydrogen over heated iron(II) diiron(III) oxide. Collect water in a U tube kept cool by immersing in cold water. (*Revise Chemistry* O-level guide, Fig. 21.2.) (*c*) (*i*) The reaction is reversible. Any build-up of products brings the reverse reaction into operation. (*ii*) Dynamic equilibrium or just equilibrium (*iii*) The forward reaction is exothermic. Raising the temperature moves the equilibrium to the left. A new equilibrium is established containing larger concentrations of iron and steam and smaller concentrations of iron(II) diiron(III) oxide and hydrogen.

22 Sulphur

1 B 2 E 3 D 4 B 5 C 6 C 7 A 8 A 9 D 10 C 11 A 12 D 13 A

23 Sulphides and hydrogen sulphide

1 B 2 D 3 E 4 A 5 A 6 D 7 C 8 B 9 C 10 D 11 A 12 E 13 A
14 A

24 Oxides of sulphur

1 D 2 B 3 D 4 B 5 D 6 A 7 B 8 E 9 C 10 A 11 B 12 E 13 A
14 C 15 A 16 A

25 Sulphuric acid and sulphates

1 A 2 C 3 B 4 D 5 D 6 A 7 C 8 D 9 E 10 C 11 D 12 B 13 D
14 C 15 D 16 A 17 A 18 B 19 A 20 A

26 Nitrogen

1 C 2 B 3 D 4 D 5 B 6 A 7 D 8 A 9 D 10 B

27 Ammonia and ammonium compounds

1 B 2 A 3 C 4 D 5 C 6 A 7 D 8 B 9 D 10 E 11 C 12 D 13 B
14 D 15 C 16 D 17 A 18 A 19 C 20 B

28 Nitric acid and nitrates

1 E 2 D 3 A 4 B 5 C 6 B 7 D 8 B 9 B 10 E

29 Oxides of nitrogen

1 B **2** D **3** C **4** E **5** B **6** B **7** A **8** A **9** B **10** D

30 Agricultural chemistry

1 A **2** C **3** D **4** B **5** C **6** B **7** D **8** A **9** D **10** C

11 (*a*) Fixation is the absorbing of nitrogen from the air into the soil.
(*b*) A nitrogen B hydrogen
(*c*) B platinum
(*d*) D nitrogen monoxide
E nitrogen dioxide (*e*) nitric acid
(*f*) Y ammonium sulphate
Z ammonium nitrate
(*g*) I fractional distillation of liquid air
II Haber process III neutralization

31 Phosphorus

1 C **2** D **3** A **4** B **5** D **6** A **7** A **8** A **9** B **10** D

32 Hydrogen chloride and chlorine

1 A **2** C **3** B **4** C **5** A **6** B **7** A **8** C **9** E **10** B **11** C **12** B **13** A
14 B **15** A **16** C **17** D

18 (*a*) Concentrated sulphuric acid (*b*) Concentrated sulphuric acid (*c*) To dry the gas. (*d*) Very soluble in water and much denser than air (*e*) (*i*) Hydrogen. (*ii*) Pass the 'salt gas' from the wash bottle containing concentrated sulphuric acid through a heated combustion tube containing iron filings. Collect the hydrogen over water. (*Note:* Any unreacted 'salt gas' will dissolve in the water in the trough and will not contaminate the hydrogen.) (*iii*)'Salt gas' contains a compound of hydrogen.

(*f*) 'Salt gas' should be dissolved in water using an inverted filter funnel attached to the delivery tube and just dipping into water (see *Revise Chemistry* O-level guide, Fig. 32.2). This avoids the water 'sucking-back' up the delivery tube due to the high solubility of 'salt gas' in water.

(*g*) 'Salt gas' in water – strongly acidic. 'Salt gas' in dry methylbenzene – neutral (*h*) (*i*) Carbon dioxide (*ii*) Hydrogen

(*i*) Hydrochloric acid (*j*) 'Salt gas' in water is in the form of ions: $H^+(aq)$ and $Cl^-(aq)$. 'Salt gas' in dry methylbenzene is in the form of covalent molecules: H—Cl.

On addition of water, ionization occurs:
$H—Cl(g) + aq \rightarrow H^+(aq) + Cl^-(aq)$

33 Carbon, carbon monoxide and organic chemistry

1 B **2** B **3** D **4** C **5** E **6** A **7** C **8** C **9** A **10** E **11** E **12** D **13** A
14 C **15** C **16** B **17** D **18** A **19** E **20** C **21** D **22** A **23** B **24** D
25 B **26** C **27** A **28** D **29** B **30** A **31** A **32** B **33** D **34** B

35 (*a*) (*i*) ethane, ethene and poly(ethene)
(*ii*) starch, maltose, glucose
(*iii*) ethene
(*iv*) ethanol
(*v*) ethanoic acid
(*vi*) poly(ethene)
(*vii*) starch
(*viii*) ethyl ethanoate
(*ix*) ethanoic acid
(*x*) ethane

(*b*) The answer can be obtained from Fig. 33.2.
starch $\xrightarrow{1}$ glucose $\xrightarrow{2}$ ethanol $\xrightarrow{3}$ ethene $\xrightarrow{4}$ poly(ethene)
1 hydrolysis; 2 fermentation; 3 dehydration; 4 polymerization.

(*c*) (*i*) Crude oil (*ii*) High boiling point fractions (from fractional distillation of crude oil) are cracked by passing vapour over a heated catalyst. Ethene is one of the products of cracking. (*iii*) If the price of oil becomes very high or if oil supplies dwindle, starch could become the source of ethene. Crude oil is a fossil fuel. Starch can be easily replenished.

34 Carbon dioxide and carbonates

1 D **2** B **3** D **4** B **5** A **6** D **7** A **8** C **9** D **10** A **11** A **12** C

35 Polymers

1 A **2** B **3** C **4** C **5** C **6** D **7** E **8** A **9** D **10** A

11 (*a*) C_2H_4

(*b*) (*i*) When ethene is passed through a solution of bromine in hexane, the solution changes from red to colourless.

(*ii*)

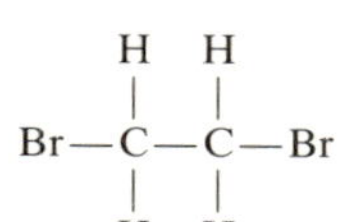

1,2 dibromoethane

(*c*) 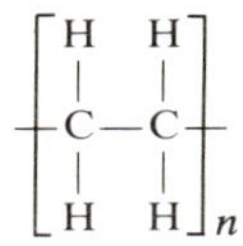

(*d*) The monomer, ethene, contains a double bond. On polymerization the double bond is removed and the polymer, poly(ethene), contains only single bonds. (*e*) Oxygen (*f*) An organometallic catalyst is a catalyst containing a compound of a metal, carbon and hydrogen. An example of an organometallic compound is lead tetraethyl, $Pb(C_2H_5)_4$, which is used to improve the anti-knock properties of petrol. (*g*) High density poly(ethene). The molecules are more closely packed which leads to the high density. (*h*) (*i*) High density poly(ethene). High melting point. It does not melt when boiling water is poured into it. (*ii*) Low density poly(ethene). Cheaper to produce. (*iii*) High density poly(ethene). Great strength and rigidity.

(*i*) 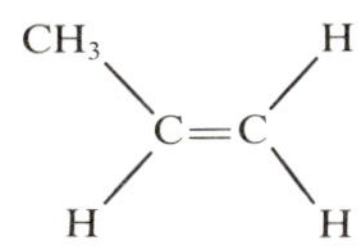

36 Fuels

1 D **2** B **3** C **4** A **5** B **6** E **7** D **8** C **9** A **10** C

37 Energy in chemistry

1 B **2** B **3** B **4** C **5** C **6** D **7** C **8** D **9** C **10** D

11 (*a*) $HCl(aq) + NaOH(aq) \rightarrow NaCl(aq) + H_2O(l)$
$H^+(aq) + OH^-(aq) \rightarrow H_2O(l)$
(*b*) Poly(phenylethene) is a very good insulator and does not lose heat to the surroundings. It also has negligible heat capacity. (*c*) 13.0°C (*d*) 1000 cm^3 (*e*) 1000 g
(*f*) Heat evolved = mass × specific heat capacity × temperature change

$$= \frac{1000}{1000} \times 4.2 \times 13.0 \text{ kJ}$$
$$= 54.6 \text{ kJ}$$

(ΔH is negative as heat is evolved.)
(*g*) Heat losses during the experiment

38 Social and economic considerations

1 B **2** A **3** B **4** D **5** E **6** B **7** E **8** B **9** C **10** D

39 Radioactivity

1 B **2** D **3** C **4** D **5** C **6** B **7** A **8** D **9** A **10** E

11 (*a*) Use no radioactive isotope. Set the counter operating for 10 minutes. Divide the count obtained by 10 to find the average background count.
(*b*) 532 − 20 = 512 cpm
(*c*) Graph
(*d*) ${}^{212}_{83}Bi \xrightarrow{\beta} X \xrightarrow{\alpha} Y$

	Element	*Mass number*	*Atomic number*
X	Polonium, Po	212	84
Y	Lead, Pb	208	82

40 Quantitative volumetric chemistry

1 C **2** C **3** D **4** E **5** D **6** C **7** D **8** B **9** C **10** D

11 (*a*) Pipette
(*b*) Indicator (methyl orange) changes colour
(*c*) $Na_2CO_3(aq) + 2HCl(aq) \rightarrow 2NaCl(aq) + H_2O(l) + CO_2(g)$
Note: If phenolphthalein is used as indicator, the indicator changes colour when the reaction takes place.
$Na_2CO_3(aq) + HCl(aq) \rightarrow NaHCO_3(aq) + NaCl(aq)$
The volume of acid required for this process is half the volume required when using methyl orange as indicator.
(*d*) 20.0 cm^3 of 0.25 M hydrochloric acid.
Number of moles in 1000 cm^3 of 0.25 M acid = 0.25
Number of moles in 20 cm^3 of 0.25 M acid $= \frac{0.25 \times 20}{1000} = 0.005$ mole

(*e*) 0.005 mole of acid reacts with 25 cm^3 of sodium carbonate solution.

$\frac{0.005 \times 1000}{25}$ moles of acid react with 1000 cm^3 of sodium carbonate solution.

0.2 mole of acid reacts with 1000 cm^3 of sodium carbonate solution.

(*f*) From the equation, 2 moles of acid react with 1 mole of sodium carbonate.

∴ Number of moles of sodium carbonate in 1000 cm^3 of solution = 0.1 mole.

(*g*) Mass of sodium carbonate $= 0.1 \times M_r(Na_2CO_3)$

$= 0.1 \times 106$

$= 10.6$ g

(*h*) 10.6 g of anhydrous sodium carbonate combines with (28.6 − 10.6) g of water to produce 28.6 g of sodium carbonate crystals.

(*i*) 180 g

(*j*) 10

41 Qualitative chemistry

1 A **2** D **3** C **4** A **5** D **6** D **7** E **8** A **9** D **10** E **11** B

12 (*a*) (*i*) carbon dioxide

(*ii*) zinc chloride

(*iii*) zinc oxide

(*b*) Zinc carbonate

(*c*) Zinc carbonate neutralizes acidity in insect bites and stings.

(*d*) Crush the mineral, then roast it in air to convert it to the oxide. Heat the oxide with carbon. $ZnO(s) + C(s) \rightarrow Zn(g) + CO(g)$